Stories of Place

Uma Kothari · Maria Borovnik ·
Phoebe Everingham ·
Karen Paiva Henrique · Fiona Miller ·
Natalie Osborne · Joseph Palis ·
Sarah Wright

Stories of Place

Geographies of Meaning, Memory and Connection

palgrave
macmillan

Uma Kothari
Global Development Institute
University of Manchester
Manchester, UK

Phoebe Everingham
School of Communication, Society and Culture
Macquarie University
Sydney, NSW, Australia

Fiona Miller
School of Communication, Society and Culture
Macquarie University
Sydney, NSW, Australia

Joseph Palis
Department of Geography
University of the Philippines-Diliman
Quezon City, Philippines

Maria Borovnik
School of People, Environment and Planning
Massey University
Palmerston North, New Zealand

Karen Paiva Henrique
Human Geography, Planning and International Development
University of Amsterdam
Amsterdam, The Netherlands

Natalie Osborne
Engineering and Built Environment
Griffith University
Nathan, QLD, Australia

Sarah Wright
Geography and Environmental Studies
University of Newcastle Australia
Callaghan, NSW, Australia

ISBN 978-981-95-4920-7 ISBN 978-981-95-4921-4 (eBook)
https://doi.org/10.1007/978-981-95-4921-4

This Palgrave Macmillan imprint is published by the registered company Springer Nature Singapore Pte Ltd.
The registered company address is: 152 Beach Road, #21-01/04 Gateway East, Singapore 189721, Singapore

To the human and non-human families that nourish us and to the places that have given us the memories we share in our stories.

Acknowledgements Those of us living on unceded lands wish to acknowledge the rightful owners of these places: Tangata Whenua o Aotearoa, the Jagera and Turrbal Peoples of Magan-djin (brisbane), Awabakal (newcastle), Gumbaynggirr people and Country, the Dharug, Garigal and Gadigal Wangal peoples and Country (sydney).

We are grateful for the support and encouragement of numerous friends and colleagues: Dominique Amorsolo, April Bennett, Tony Birch, David Bissell, David Brown, Kay Hammond, Anne Hiha, Shelley Hoani, Donna Houston, Lisa Palmer, Emily Potter, Jo Sharpe, Aliitasi Su'a-Tavila, Katerina Teaiwa, the Brisbane Free University Radical Reading Group, GeonarrativesPH at the University of the Philippines-Diliman, members of the Living Places Centre for Environmental and Spatial Justice at the University of Newcastle, the Media, Literary Geographies and Geohumanities (MELANGE) Research Cluster-UP Diliman Department of Geography, the Narrative Ecologies of the Warragamba Dam ARC Discovery team, members of the Shadow Places Network who share a love of stories, and the Women Writing Away work in progress groups.

The writing of this book was supported through funding received from the Manchester-Melbourne Strategic Partnership and the Journal of Development Studies-Global Development Institute conference support fund to hold two international workshops on Storying Geography at the University of Melbourne in 2023 and in London in 2024. Additional funding to facilitate the participation of members of the Storying Geography Collective at these workshops was received from the University of Newcastle, Macquarie University, the Australian Research Council (ARC) Future Fellowship FT160100353 and the University of the Philippines-Diliman. This volume was made open-access thanks to the financial support from the Amsterdam Institute for Social Science Research (AISSR) at the University of Amsterdam.

Competing Interests The authors have no competing interests to declare that are relevant to the content of this manuscript.

Contents

About the Authors

Uma Kothari is Professor of Migration and Postcolonial Studies at the Global Development Institute, University of Manchester, UK and Honorary Professor of Human Geography, School of Geography, University of Melbourne, Australia. Her research interests include colonial legacies and decoloniality, migration and mobilities, postcolonial travel and island geographies. She is currently a Leverhulme Major Research Fellow on the project 'Travelling Imaginaries: the adventures of postcolonial travellers in the 1950s'.

Maria Borovnik is a Senior Lecturer in Development Studies, School of People, Environment and Planning, Massey University, Aotearoa New Zealand. Her current research interests lie in the intersection of mobilities, development, and maritime geographies. She is currently involved in research exploring how seafarers have dealt with the global crew change crisis during COVID-19, which is supported by Te Apārangi - The Royal Society of New Zealand Marsden Fund.

Phoebe Everingham is a Lecturer in Environment and Society of Geography and Planning at Macquarie University. Her research focuses on the intersections between tourism development and sustainability with particular emphasis on regenerative economies, post capitalism and decolonial futures.

Karen Paiva Henrique is a tenured Assistant Professor in the Department of Human Geography, Planning and International Development at the University of Amsterdam. She is an urban and feminist political ecologist with a multidisciplinary training in geography, planning and design. Her current work investigates the un/just implications of climate adaptation, the role of storytelling and knowledge exchanges in addressing urban crises, and everyday decision-making in the context of climate change.

Fiona Miller is Associate Professor of Geography and Planning at Macquarie University. She conducts research from a political ecology perspective on the social equity dimensions of environmental change, climate change vulnerability and adaptation, and society-water relations. Fiona is currently undertaking research on place-based practices of repair, community experiences of floods, and climate-related loss and displacement.

Natalie Osborne is a Senior Lecturer in urban and environmental planning in the School of Engineering and Built Environment, Griffith University, on unceded Jagera and Turrbal Country. Her research focuses on emotional geographies, public feelings, and social, spatial and climate justice. She is a co-producer of the *Radio Reversal* radio program and podcast, and an organiser with the Brisbane Free University.

Joseph Palis is Associate Professor at the University of the Philippines-Diliman Department of Geography. He teaches cultural geography, geohumanities, countercartographies, and island and archipelagic geographies. He heads the GeonarrativesPH Collective at UP Diliman.

Sarah Wright is Professor of Geography and Development Studies at the University of Newcastle, Australia. Her research focuses on critical development studies, geographies of weather, and Indigenous/postcolonial geographies. She collaborates with Indigenous-led research collectives in Australia and place-based social movements from the Philippines and throughout Southeast Asia.

List of Figures

1

Positioning: Stories of Place

Storying Geography Collective

Abstract There is intensified interest in stories and storying across many social sciences, and in how stories may offer and reflect alternative ways of knowing and encountering place. In this introductory chapter, the *Storying Geography Collective* considers the ways in which stories of place *matter* and outlines some of the key issues and theories we engage with across this volume. We offer an overview of existing literature from geography and across the social sciences and humanities that engage with the importance of stories and storytelling, particularly in matters relating to environmental and climate justice, more-than-human and place agencies, emotions and embodiment, material struggles, social and epistemological justice, the sovereignty of First Nations Peoples, and transformative politics. We contend with how stories can reinforce existing, oppressive power structures, and dominant ways of being and knowing, and the ethical complexities shaping who can—or should—tell which stories, of where. In *Positioning*, we locate our Collective in this work of understanding how place and stories matter to each other, and chart an approach to stories for this volume that is attuned to their possibilities and affordances, while attending to the ways power is at work in stories, and the ethical responsibilities of storying place

U. Kothari et al., *Stories of Place*, https://doi.org/10.1007/978-981-95-4921-4_1

Keywords Storytelling · Storying · Politics · Ethics · Methods

We gather in an old house on the edge of Hampstead Heath in London to share our individual stories of place. As each of us reads our story out loud we listen carefully, revelling in the resonances that emerge between our experiences and emotions. The stories reveal multiple connections between memories and sense making and uncover intimate relations between more-than-humans and places. Our stories bring to the fore tragedy through colonialism, loss and grief, as well as a sense of playfulness and delight. Through their rhythms and cadences, the stories resist spatial and temporal fixity. They are tales of searching for places, of places that cannot be entirely remembered or fully grasped, of the expansiveness and confinement of places, of new places and of ones that have changed and transformed us, over time. We become aware of the vibrancy of places, including this place here by the Heath, how they hold so many different meanings and imaginaries and can elicit such distinctive feelings. Our stories invoke sensory qualities, of colour, shade, smell and touch, that linger beyond the story and that stimulate affective responses of fun, sadness, anger, grief, abandonment, enchantment and pleasure. We are altered through these moments of storying together and as we scatter and return to our homes, we carry with us a renewed sense of how deeply place matters.

Stories Matter

This book is about place and stories. Stories matter; they are central to the ways we produce and share knowledge, make meaning and communicate the world. As Ben Okri reminds us, "anything that affects our perception of the world usually comes in the form of a story" (2015, p. 1). Stories help us narrate our lives with and in place, bringing ourselves and the diverse places of our lives into relationship. Yet, when thinking about stories, place is often overlooked, particularly within dominant Western traditions, relegated to a backdrop or context, sometimes insubstantial, irrelevant or substitutable. But place has agency and meaning beyond humans, and is central to storying as inspiration, as guide, and as a vital source of connection and identity. In this book, we explore the relationships between stories and place to reveal how stories come from, are made by and make different kinds of places and how places themselves have agency to make and recount stories.

There is an intensified interest across the social sciences in both place and stories. As geographers, the concept of place is central to our work (Edensor et al., 2020; Massey, 2013; Tuan, 1975). Beyond an understanding of place as a point in space, a specific locality, geographers have long been interested in how culture, identity, emotions and politics shape relations with place. Recognition that places are not fixed or bounded in time and space, but shaped by material and cultural processes, everyday practices and networks of relations, has led to an interest in diverse experiences and understandings of place (Cresswell, 2014). A sense of place and attachments to place have come to be of particular interest in contemporary geography as the intensity of global economic and environmental changes transform human (and more-than-human) relations with place. Tending to stories, while vital in geography, has been less central. A relatively recent turn towards stories and storytelling in, for example, work on landscapes (Smith & Fiore, 2010), climate change (Harris, 2020, 2024), literature (Hones, 2022), environmental humanities (Hughes & Moscardo, 2024; Stadler et al., 2016) and research methods and fieldwork (Gegeo & Watson-Gegeo, 2001; Smith, 2014), has emerged through increased attention on relationships in the discipline, and a focus on embodiment, creativity and affect including through work in non-representational geographies (Daniels & Lorimer, 2012; De Leeuw & Hawkins, 2017; Lorimer, 2003; Rose, 2016; Springgay & Truman, 2018; Vannini, 2015). For other disciplines with a long tradition of, and interest in, stories and storytelling, such as literary studies (Hamilton et al., 2024), history (Cronon, 1992), anthropology (Cruikshank, 1999; Götsch & Palmberger, 2022; Jackson, 2002), philosophy (Latour, 2014) and environmental humanities (Houston, 2013; Houston & Vasudevan, 2017; Van Dooren & Bird Rose, 2016), the recognition of the importance, and agency, of place has brought attention to matters of the more-than-human, environmental crises, environmental justice, borders and sovereignty.

This coming together of stories and place is both timely and exciting. Our work as the Storying Geography Collective seeks to contribute to these emergent discussions and to speak to the potential of place to marshal intimate and embodied knowledges that respond to place affectively. We look to the ways stories allow us to centre trauma, pain, loss, love, hope, nostalgia, melancholia or other myriad emotions and relationships. Stories connect us to place, are told in, with, and through place yet leave space for uncertainty, contingency, interpretation and incompleteness (Masud, 2023). And, stories, too, make place with their intimate tellings (see Bondi, 2023). Stories told or written "from the body" of a storyteller, who temporarily (or repetitively) engages with their "material experience" of a place (Bridger et al., 2022, p. 157, referring to Françoise Defroment in Wilcox et al., 1990), bring together the interaction between storyteller and place. As such, stories can connect the territory of the external world with our memories, and in the process add to our perceptual 'maps' (Bawaka Country et al., 2022; Benjamin 2019; Bridger et al., 2022; Morgan, 2019).

Yet, stories are by no means unproblematically progressive. While they can comfort, offering a sense of familiarity and escapism, they can also alarm. Stories can relate to and illuminate places of oppression, such as the camp, states of exception and the colony, as well as propel their creation. Narratives of place can also reinforce processes of displacement, dispossession, alienation and extraction underpinning material forms of political economic power. Thus, telling stories, and maintaining or imposing silences, is never innocent but always politically and ethically fraught.

Our attention to stories and place in this book and through our Collective aims to illuminate every day, mundane experiences of, and practices, in place. Yet, we also seek to speak to big stories, to ethics and politics, in ways that recognise the 'big' and the 'small' as deeply entwined. Lorimer suggests that it is the small stories that allow us to "explore the fabric of everyday life" (2003). However, while often considered such, feminist scholars and others insist that personal, biographical and everyday stories are not so small. Rather, through stories, everyday places, experiences of place and being in place, can be brought to the fore, understood for their richness and significance, and speak to broader scales, structural processes, oppressions and resistances. As Vasudevan et al. (2023, p. 1728) argue, "Anticolonial feminist storytelling alters the spatiotemporal scales through which planetary crises are understood by centring the relationship between body and land". New possibilities, insights and capacities to imagine otherwise can be generated through engaging with these big and small stories of place. Thus, stories live not only in the everyday, but also across different temporalities and scales and are always imbued with power relations. Just as global dynamics live in the microscales of embodiment and interpersonal relations, those oft-dismissed and intimate stories of home and body speak to matters of profound consequence. As Ross Gibson (2010, p. 10) writes, while our stories can "start and stay in privacy we can also use them to pronounce to and on the larger world". Through connecting different stories and telling stories together in ways that reveal shared experiences, our mundane realities can create grander narratives connecting the body, home, landscape, nation and the larger world.

In this volume, we are concerned with attending to stories in ways that centre their complexity, their world-making potential, their ethics and their fraught potentialities. We do this through a series of chapters that engage with stories in an expository sense while also sharing stories of place, in its varied forms, emotional entanglements and understandings. The collection is structured around the themes of Responding, Imagining, Haunting, Situating, Relating, Journeying, Dwelling and Enchanting. Through these themes, we address questions relating to what work multiple, diverse and varied stories do in storying place, what are the political implications of the stories that are told, shared, heard and amplified, how do places themselves tell stories, and what could be the processes by which stories can be co-created by the human and more-than-human. As we seek creative ways to carefully navigate this terrain, we pay attention to power and positionality, and specifically to how stories might provide an alternative lens 'to pronounce to and on the larger world'. Our chapters reveal how storying can be a practice of hope, albeit one that recognises injustices of the past, without denying pain or loss, and how stories might move us towards an accountable responsibility to place, to each other and the living world.

Returning to Ben Okri, we seek inspiration in his elaboration of stories as, "the infinite seeds that we have brought with us through the millennia of walking the dust of the earth … They are whispers from beyond" (Okri, 2015b). These are some of the whispers we listen to; this is some of the dust we seek. Stories enable us to travel through time and space, to learn from the past, engage with multiple presents and imagine different futures. While stories have the potential to foreclose, they can also create open, progressive futures through acknowledging different pasts and manifold presents. It matters, after all, "what stories we tell to tell other stories with" (Haraway, 2016, p. 12). Keeping these matters in mind, throughout this book we ask: How and in what forms might stories nurture co-existence, unfold progressive insights and deepen relationships with and in place?

Politics of Stories and Storytelling

Drawing upon, and creating, diverse experiences, ideologies and forms of knowledge, stories are always and inevitably political, shaping who belongs in, and who is excluded from, place. Relational entanglements between stories and place are therefore complex and fraught. Reflecting dominant power relations, stories and storytelling can create inequalities and forms of exclusion as well as incite injustice and violence. They can provide a platform for prevailing voices, rendering them sympathetic and legitimate, while silencing and compromising the stories of those they oppress (McQuire, 2024; Wright, 2016). Some stories then can reinforce, rather than challenge, dominant subjectivities, positionalities and western epistemic privilege, and reinscribe rather than dismantle binaries between 'emotion-reason and people-place' (Ortiz, 2023). Thus, not all stories or practices of storying are progressive, inherently liberatory (Ortiz, 2023) or uniquely innocent.

If stories are potentially a powerful means to create and share knowledge about places that shape our modes of being, we must also contend with how they can be put in the service of coloniality, empire, extraction and colonial racial capitalism (Banivanua Mar, 2012; Birch, 1992; Watego, 2021). Moana Jackson reminds us that colonisation involved the brutal taking of Indigenous peoples' lands and lives, processes framed, rationalised and justified through racist, pseudo-scientific and legal accounts (2002, p. 134). Examples of the potential danger of stories that are being told by those who have dominated narratives over centuries and are continuing their "denial of plurality" as Kothari and Klein (2023, p. 43) put it, can be found in a settler context, such as in so-called australia. Note that we have chosen to resist this European-flavoured term in using small lettering. We are keen to emphasise how the role of stories in settler-colonies, and other colonial contexts, can uphold and maintain colonial possession and other forms of violence.

Stories have been used by and for the powerful to romanticise, launder, naturalise and justify systems and acts of domination, exploitation and violence. The shape and content of these narratives vary according to context and function. For example, whitewashed stories of history are part of the mental and affective infrastructures that sustain settler colonialism, imperialism, white possession and supremacy (Banivanua Mar, 2012; Behrendt, 2006; Birch, 1992; Watego, 2021). In australia, stories about 'heroic' imperial figures, representing pioneers fighting to survive 'in a harsh and empty land' (Birch, 1992), support the fiction of *terra nullius* on which australia is founded, and which continues to underpin its imagined legitimacy and ongoing existence (Watego, 2021). These stories feed a shared colonial imaginary and provide a narrative framework that justifies racist policy frameworks and state violence. In this context, Amy McQuire (2024), in her account of the Palm Island uprising—a small island off the north-east coast of australia—explored how Islander protestors were narratively represented as being armed with 'imaginary spears' to story them as violent and menacing, in turn 'justifying' police and state violence and repression. McQuire (2024, p. 25) writes: "The spears were not there, but their existence did not matter because the stories filled the gap". Tracy Banivanua Mar (2012) provides another example of the violence of settler storying in which stories of (actual) massacres by white settlers against the Juipera People of the place now known as Mackay (north-eastern australia) served as both an "extinction narrative" of the Juipera People, and an "origin story" for the region's settlers (Banivanua Mar, 2012, p. 187). This narrative infrastructure is used to refuse Juipera People's contemporary claims for justice, reparations and return of stolen land—because according to (self-serving and erroneous) settler stories, nobody is left.

Other political projects use stories to create structures of feeling that sustain violent imaginaries and justify genocide. Fascist discourses, for instance, continue to rely on stories—including those entangled with folklore, sci-fi and fantasy—to propagate ideologies and imaginaries, creating shared affects that can be put to political use, laundering violent projects into something that is more broadly acceptable (Freistein et al., 2022; Gomel, 2000). 'Official' stories are also used to embed, delimit and reify regressive narratives about people and places, producing singular meanings that come to dominate. This is evident when public statues, plaques and memorials, for example, narrate particular histories and valorise certain figures, shaping and transforming places and our relationships to them (Carlson & Farrelly, 2023). Clearly, there is important work to be done in "question[ing] the truth of our own and others' stories … to be attentive to how power shapes which stories get told, get heard, and carry weight" (Sandercock, 2003, p. 12), and how dominant storytellers wield their stories violently, to speak over and against the stories, storytelling and existence of others (Wright, 2016). Such accounts propel and justify particular relationships with place, bringing into being spatial realities through creating places in ways that exclude some and include others.

Contested Narratives

Dominant narratives, however, have always been contested, with counter-stories and stories of resistance challenging, defying and refusing these accounts. Storytelling is, as Vasudevan et al. (2023, p. 1730) write, "a fertile ground of struggle". Recounting and telling stories can be empowering, can subvert the systematic silencing of the marginalised and ignored, and can create a space to reclaim personhood and bring about political change. Indigenous and other marginalised peoples have spoken back against racist and colonising stories and have reclaimed others that have long been suppressed and silenced (Miles, 2019). Engaging with counter-stories helps us attend to what is not included, and what is mis-storied, in historical records and official tellings. Through processes of decolonisation, for example, colonialist stories about places are critiqued, dismantled and replaced with new and reclaimed ones. Katerina Teaiwa's (2015) work on phosphate imperialism contrasts dominant stories in australia and Aotearoa that erase the impacts of phosphate mining and exhaustion on Banaba Island and the ways these practices sustained settler communities in the Pacific. She contrasts this with narratives that name and call out this absence, to demonstrate the importance of listening to the stories of people of the land and sea. Such counter-storying draws clear connections between seemingly distant places, explicating relations of responsibility for harm and injustice. These differently storied histories of place point to absences in the archives and the violences and lived realities obscured by dominant accounts (Trouillot, 1995), that are captured in the stories of war and confrontation between settlers and Aboriginal people across australia (see Spearim, 2025 Frontier War Stories podcast). Counter-stories thus form an essential element of truth telling towards restorative justice. They have the potential to powerfully remake and revise, recounting the past and present differently and enabling alternative possible futures in a context in which "rival narratives of past, present and future are constantly in contention" (Hassett, 2024, p. 151).

Importantly, stories have the potential to reveal different understandings of place—what it is, what it means, what it does and what it could be. For many Aboriginal people in australia, for example, stories are relationally produced by and with Country. Country is the lands, waters, skies and myriad beings, emotions, dreams, processes and people that make up place. Here, Country and its beings—including animals, birds, winds—make stories; they communicate, guide and create (Bawaka Country et al., 2022; Wright et al., 2009). Place can be agential and knowing; it can hold and tell stories. And stories make the land itself. Additionally, as Somerville et al. (2023) illuminate, processes and outcomes of active learning can emerge through collective place-based storying in ways that connections to Country and place can be collectively negotiated. Engaging with different understandings of place is a core concern in geography, with stories an important method by which these can be facilitated.

In imagining otherwise, and giving shape to other worlds, stories can resist the apparent inevitability of dominant politics and ways of engaging with place. Through their critique of the whiteness of australian law schools and the erasure of Aboriginal knowledge in australian legal education systems, Watson (2024) demonstrates the power of counter-storytelling and the stories of 'outsiders' who do not belong to the dominant system, to expose inequalities and challenge their portrayal as immutable and inescapable. Drawing on Delgado (2013), Watson (2024, p. 197) writes that outsider storytelling is a "powerful means for destroying mind-set—the bundle of presuppositions, received wisdoms, and shared understandings", through which dominant discourses are framed. Counter-stories are thus part of a prefigurative politics that can bring to light other realities that have been overshadowed or deliberately excluded. They challenge the dominance of particular stories of place, such as that of the colonial fiction of *terra nullius*, and through their telling enact more hopeful politics of resistance. Vasudevan et al. (2023, p. 1) argue that revealing dominant stories of oppression and extraction to be a fiction of modernity can "open a space of possibility, to tell stories otherwise that reimagine what it means to be human on earth". These authors argue for '[s]torytelling as anticolonial praxis'. In so doing, counter-stories of place can act to decentre and destabilise apparent 'truths', creating opportunities to understand other relations with, and understandings of, place that have been deliberately denied and even violently destroyed. It is only when counter-stories become dominant that a more progressive politics with place can be realised and more just futures and relations storied, envisaged and actualised. Yet, for this to occur audiences need to listen, to question their assumed knowledge and be open to relating to place differently.

The prevalence of dominant narratives, on the one hand, and the vibrant realities of counter-storying as struggle, on the other, requires thoughtful consideration of how stories and storytelling can be pursued in ways that avoid replicating the politics of oppression and marginalisation, and that do not silence or otherwise exclude diverse relations with, and understandings of, place. As such, how can stories be told differently, and more ethically, in ways that do not engage with place in extractive and commodified modes? How can an appreciation of counter-stories and storytelling help resist imaginative, discursive and material processes that contribute to, launder and justify exploitation, extraction and dispossession? How can the sharing and telling of stories, particularly by those with power and privilege, avoid reproducing processes of dispossession and extraction? We know that to be well-meaning is not enough.

These questions and concerns motivate the work of our Collective. Yet, of course, there are no easy answers to how to navigate these terrains. In our work, we have looked to ethics to help guide how we engage with stories to inform our research, our praxes and our modes of being.

Ethics of Stories and Storytelling

In this book, we recognise the importance of approaching stories respectfully. This includes acknowledging the diverse and long-standing intellectual traditions of storytelling, recognising prior belongings and storytelling traditions, attending to silences and refusals, and honouring the agency of places.

While storytelling and storying share features with other forms of geographical writing and inquiry, they differ in approach, function and emphasis. The mode of storying we describe here does not necessarily begin with a research question nor does the story always serve as a vehicle for analysis. Instead, we start with a story and generate meanings with it. Storying is not so concerned with producing new data or narrativising a data set, but rather offers a way of reorienting ourselves to what is often already known and felt. In this way, stories create meanings and connections through re-encountering and re-making memories of lived experience, the everyday and the familiar. Furthermore, storytelling, while not unique in geographical research in how it attends to the positionality of the author, emotion and affect, the relationship between storyteller and place, and the connections between storyteller, reader and listener, there are differences in emphasis. The telling and the listening of a story require responsiveness, a willingness to move and be moved.

Thus, storying places greater emphasis on cultivating this kind of affective engagement that is not always so readily elicited in other forms of geographical writing.

Stories and storytelling are, of course, not new, and the intellectual traditions that hold them are rich, plural and deep. As such, an important responsibility for our work, as a Collective, is to engage with Indigenous, decolonial and feminist theorists (De Leeuw and Hawkins 2017; Keeshig-Tobias, 1990; Smith et al., 2021; Todd, 2016), who point to some of the different lineages, co-becomings, knowledge traditions and processes of meaning-making in and with place. For many Indigenous scholars, place, land and Country have always had agency, telling stories that bring people into relationship. Those marginalised by mainstream knowledge practices have long turned to stories to theorise life, place and community, and to tell and share stories to critique and resist dominant narratives. We attend to stories to complement some of the ways that analysis and meaning-making can be carried out in spaces beyond dominant practices of academic knowledge. We are inspired by work on citational politics and relationality (Liboiron, 2022; Mott & Cockayne, 2017; Tynan & Bishop, 2023), and seek to engage with marginalised perspectives on, and stories of, place.

We hope to avoid replicating extractive and exploitative relations with places and peoples and the more-than-human world common in many forms of academic research. This is not a straightforward task. Stories can inadvertently transgress moral and ethical boundaries. Stories can be bent, simplified, obfuscated or manipulated (Meretoja & Davis, 2017), and misinterpreted, perhaps especially when removed from context and the relationships that help give stories meaning. Storying a place could (mis)represent, simplifying versions of a place, flattening it out to make it 'knowable' and be 'mastered' by the storyteller. Instead, storytelling requires "careful, nuanced, multi-layered approaches" (Hancox, 2021, p. 117). A deep consideration of which places are being storied and whether the stories and voices may belong to someone or something other than the storyteller is an important starting point. Places are intricate, filled with agency and soaked with different histories of exploitation, hurt and harm, and of relationships, connections and joys that are not always ours to tell.

The complexities involved in telling stories about a place include relationships with human and non-human 'others', who may have lived and died a long time ago, who are alive but absent from the process of storytelling, or who have some kind of investment in, and relationship with, the places of a story. This is particularly so in a settler context, or on stolen land, like Country (of so-called australia), or land and environmental features that belong to tangata whenua (people of the land) by Māori in Aotearoa New Zealand. Stories, unlike some other forms of knowledge, may be better equipped to embrace such uncertainties, with careful listening, hearing (Frenette, 2024) and witnessing (Wright, 2022). Indeed, these relational processes may lead us away from the possibility of certainty at all; the longer we listen, hear and engage with place, the more we might find it impossible to understand what is actually going on, "until the whole place becomes strange" (Perec, 1997, quoted in Cresswell, 2019, p. 5). It is through this experience of uncertainty, and indeed disorientation, that it is possible for us to access the humility required to listen deeply and genuinely reflect on relations with place.

In engaging with stories of place, it is important to attend to silences, refusals, and that which is hidden. A story might have the potential to uncover the invisible when, as Donna Hancox (2021) explained, it is told in a way that those who own the story want it to be told. Storytellers have the power to make decisions on what to include and what to leave out; this can be both liberatory and extractive. Further, some stories should only be shared selectively, within certain relationships or spheres of belonging, or only in certain ways and by certain people, or, indeed, not at all (Carter, 2006; Moss & Thomas, 2021). Where some silences must be broken as they may represent colonial oppression, some should not (Carter, 2006; Moss & Thomas, 2021). Thus, when seeking to tell a story that is not our own, we must be careful to respect these refusals and silences. Some stories are simply not for all to hear, or know, or tell, and may be intentionally concealed in order to protect the story, and those it concerns, from the impacts of misappropriation, misinterpretation, weaponisation and theft (Carrasco, 2025; Kanngieser et al., 2024; Rubis & Theriault, 2020).

It is important, then, to be aware of our individual perspectives and appreciate the influence and agency of those who have shaped places over time. Yet, as Meretoja and Davis (2017, p. 8) find, even with best intentions, "our moral vision may be flawed". Recognising these limitations, we acknowledge how our relationships with place and the stories we tell emerge from our own positionalities. We acknowledge prior connections, stories and belongings, and as these resonate with our own positionalities, the voices of places we story about become entangled in a creative network of relationships.

Before working with human participants, our academic institutions require ethical applications, usually requiring researchers to consider matters around confidentiality, privacy, access to and use of information. More so, we are asked to reflect on any gender and cultural concerns, reciprocity and relationality, and how participants' dignity will be respected throughout the process. Similar ethical issues need to be addressed when our stories involve more-than-human participants, or when storytelling is our method of engagement, because "there are no ethically neutral narratives, and the ethical dimension of a narrative takes shape in dialogue" (Meretoja & David, 2017, p. 7). An example of how this ethical approach to storying with place worked in practice is that of the Bawaka Country authorship group (2013, p. 77), who acknowledge "Countries as legitimate sites of knowledge authority", and highlight Yolŋu authors as the "true knowledge authorities" on Bawaka Country: because they are "part" of it. These Indigenous connections between Country, landscape features and places are recognised as interconnected authorities. As Yandaarra (Country authorship), led by Aunty Shaa Smith points out, if more-than-human agency is taken seriously, so too must issues of respect and consent with more-than-human agents, particularly with Country (Yandaarra with Gumbaynggirr Country including Aunty Shaa Smith 2022).

Slowly, there is increasing recognition of the rights of nature, for example, in Bolivia and Ecuador where "Mother Earth" is recognised as sacred and interrelated with all living beings (Crews, 2019, p. 315), and in the recognition of the Whanganui river as possessing the right "as a legal person", in Aotearoa New Zealand (Youatt, 2017, p. 49). Even in built over places, where nature is "made to run in dark spaces", "the sacred still swims; it still moves and stirs beneath us, reminding us of all that feeds" (Case, 2021, p. 135). This was in reference to Wellington's Kumutoto stream, "most of which flows underground, under layers of concrete and colonisation, on a now-curated path under roadways and city footpaths before emptying at the ocean" (Case, 2021, p. 135). When storying about other built places, then consideration must be given to what kind of sacred land they may cover or obfuscate, but which remains.

In this volume, having storied together, we have been guided by five specific ethical questions. First, whose voices are represented when telling stories, and how do we respectfully listen when engaging with place? Secondly, in what ways are these voices and our storytellers' positionalities intermingled, and what obligations of accountability and reciprocity arise from these relationships? Thirdly, when bringing the audience into these relationships, how might the story connect with that audience and be understood, and what cultural values, relationships, differences or concerns need to be considered? Fourthly, what are the ethical issues relating to ownership of and access to the places in our stories, and what should be left unsaid in order to respect the privacy, authority and autonomy of places and their caretakers? Finally, where is the potential for harm in the telling of these stories of place and what possibilities for different modes of thinking about and being with place, more ethically and responsibly, might emerge from storying in these ways?

Aware of our responsibility to write ethically and transparently, and mindful of our positionality, in the following section, we discuss the methods we have used in developing the stories included in this volume, and how working as a Collective shaped our practices of storytelling and meaning-making. We acknowledge that there are many ways of telling stories that serve different goals (Bawaka Country et al., 2022; Bönisch-Brednich & Meyer, 2024; Cameron, 2012; Wright et al., 2012). Here we document our own.

Approaching Storying

This book considers the potential of stories to illuminate and amplify places, relationships and perspectives that have been excluded, hidden and overlooked. It is dedicated to disentangling the power, ethics and political implications of storying place in recounting and (re)shaping worlds.

This collection centres creativity, to explore the possibilities and affordances of stories and storying, to help understand varied and plural storytelling and to practice relational epistemologies. As Cameron (2012, p. 573) writes, "story draws attention to the relationship between personal experience and expression, and the broader contexts within which such experiences are ordered, performed, interpreted, and disciplined". What was crucial, for us, was doing this work collectively, as a way to support experimentation and to practice some of the ethical considerations outlined above.

We first came together in 2020 to share our individual experiences of the unfolding global pandemic. Some of us had prior connections and relationships, some of us were, at that point, strangers, drawn together by virtue of all responding to an expression of interest. Through our conversations, we began to foreground storytelling as a tool and process to develop ethical scholarship, while being attentive to mounting (and often gendered) responsibilities at home and at work. As a starting exercise, we each wrote a vignette, telling a facet of our COVID-changed realities, and coming together via Zoom to share these stories, reflect on their meanings and map their interconnections as well as points of divergence and departure. We brought our stories together alongside describing the process of storying collectively in the article, "Storying *Pandemia* Collectively" (Storying Geography Collective, et al., 2023). Crucially, this work allowed us to make sense of the emotional and material turmoil that surrounded the COVID-pandemia, and heightened our collective interest in the power and role of stories and storying. There are so many ways that stories can be told and through many different media. We chose to focus on the written. Yet, as the full vignettes could not be included in the article, we each audio recorded our story in full. The different voices, inflections, accents, cadences and paces added a richness to our stories. In narrating our stories, we engaged in what MacBeath calls the "articulation of [our] own inner voices" (Ahmed 2006, p. 196).

The work documented in this current volume began in early 2023. This time, we focused our initial vignettes on *place*—each choosing a place that held meaning for us. We shared our stories and collectively discussed their content, purpose and the ethics of 'storying place'. These conversations unfolded slowly. We revisited our stories several times as we took turns reading each other's aloud to the group, and considered: What do our stories and our individual methods of storytelling offer towards understandings of place? What insights can we gain from collectively reading texts? How do our stories differ, and what connects them? How does this collaborative relationship affect how we story and receive each others' stories? What differences in interpretations do we notice? How do our own individual stories of place resonate with those of others, and what are the various affective qualities of stories? What is it that matters, in these stories, and in these acts of storying?

The process of 'storying place' as a Collective has allowed us, as storytellers and listeners, to explore how places are remembered and transformed over time, foregrounding the fluidity of memory and the emotional weight of nostalgia. Each story sought to transport its audience to a remembered place. Yet, the audience's own experiences of the place being storied inevitably add layers of interpretations that can expand a story's meaning and its purpose. Considering the role of the audience has made us reflect on the political and material implications of those stories that are not told, that remain silenced. Hearing others read and interpret our stories also drew attention to the sometimes sensitive, sometimes complex, links between personal identities and shared stories, the attachments they highlighted, and how stories can be understood and mobilised beyond their original intention. Crucially, our sharing and processing stories together have made us acutely and viscerally aware of how stories are told, and how they are shaping and reshaping our worlds.

We outline this process here to document and add transparency to our project and to make visible that this has been slow, careful and respectful work, developed through critical reflection, collaboration and building relationships; characteristics not always found in traditional academic spaces. Ultimately, working with stories has created more space for vulnerability, uncertainty, humility and experimentation, and more time to nourish connections that sustained the work.

How the Book Unfolds

This book is organised around ten verbs, beginning with 'Positioning' as our introduction to situate this work in a broader landscape of scholarship on stories and to locate the Storying Geography Collective. It concludes with 'Inviting', an epilogue which offers some reflections on approaches to storying places. This is not quite an edited collection, nor is it a fully co-authored manuscript, so we would like to offer a brief explanation on our process of writing and how this book unfolds.

We collectively decided to each write a story of, and with, *place*. After circulating initial drafts, we spent time reading and engaging with each other's stories. Reading the stories together, we considered the kinds of understandings, meanings and modes of relating to, and with, place that each story offered. Insights from these discussions were often folded back into the stories as they were re-drafted and finalised, but the stories themselves were, essentially, individually written.

Chapters 2 to 9—*Responding, Imagining, Haunting, Situating, Relating, Journeying, Dwelling, Enchanting*—each include a story and an elucidation of key concepts about place that the story reflects. Part of the beauty of stories is how they allow for multiple meanings and interpretations; we did not want to foreclose that by being over prescriptive in the writings that accompany our stories. As we sketched a synopsis for each chapter, these verbs emerged and became a touchstone for each of us as we worked on our contributions.

As you read through this collection, you will see that there are a variety of ways the stories of and with place have been used. In some cases, the story is presented in full, distinct from the interpretive element. In some, the story is excerpted and woven throughout while in others, the story and reflections have, essentially, merged and are now largely indistinct from each other. We note this simply to highlight that there are a variety of ways to work with stories in academic writing, indeed more than have been exhibited here, and we hope this diversity offers encouragement to experiment.

Responding, by Fiona Miller, considers how stories of the mundane and every day, and of loss and absence, can help us understand connections between local experiences in place and global forces of injustice and harm. Via a story about potholes on a cycling route, *Responding* calls us to attend to processes that create harm, acts of resistance, repair and responsibility, and to consider how stories demonstrate what we value, what we care for, about, with.

Imagining, by Uma Kothari, navigates the complex contour lines between colonial and postcolonial imaginaries of place and attachment, migrations between colony and metropole, and the intimate process of unlearning colonial sensibilities through and with family and place. Through a story of the search for an idealised English countryside, *Imagining* speaks to how dominant colonial stories can be slowly undone, their mythology quietly unravelled as new, postcolonial stories are created.

Haunting, by Sarah Wright, offers stories from childhood. She writes of days spent playing in a cemetery, a more-than-human and multiply storied place, and the unexpected encounters and opportunities for care, listening, noticing and re-membering this place offers. *Haunting* asks us to attend to the ongoing presences of those who have departed, and to the different experiences of temporality in stories that places can offer.

Situating, by Joseph Palis, is concerned with practices and experiences of migration and mobilities that dislocate and relocate us in place, across years and borders as we depart and return. Joseph considers being lost and trying to find oneself at home again. *Situating* connects an intimate and global story to the fissures of coloniality, displacement and exile, to reveal the ways people may be divided from place, and the relationships between return, repair and reconnection.

Relating, by Maria Borovnik, emphasises embodied and sensory experiences as we encounter places through walking, attending and connecting. Maria tells a story of her time spent on the upper Tarawera River in Aotearoa (New Zealand), and considers how to ethically, with care and concern, relate to places as tauiwi (foreigner). Uncertainty and wonder shape this account of the relationships between places, stories, knowledge and sovereignty, as *Relating* foregrounds some of the important reflective work that makes relationalities possible.

Journeying, by Karen Paiva Henrique, attends to mundane and bureaucratic spaces, liminal spaces encountered when on the move. She shows how these places become imbued with sensations and significance. Drawing on Latin American and feminist scholarship that speaks to exclusion and marginality, Karen considers (non)belonging as an active and often collective process of meaning-making with places.

Dwelling, by Natalie Osborne, considers the politics of dwelling as a settler on stolen land, and how inhabitation is both enabled and troubled by places and non-human Others. This story of a dilapidated and beloved timber house, where Natalie dwells in collaborative and contested community with plants, critters, fungi and floodwater, confronts the myth of the bordered and bounded self, and settler-colonial-cis-heteropatriarchal notions of mastery and control over the domestic space, and seeks out possibilities for dwelling otherwise.

Enchanting, by Phoebe Everingham, transports us to the Sandhills Community garden on unceded Awabakal land in Moolimbimba/newcastle (australia). Phoebe's narrative illustrates the multi-layered, co-existing and co-constituting histories and meanings of this place, and the manifold attachments people and non-humans have to it. Attending to contestations, violence, joy, care, grief, complexity and wonder, *Enchanting* demonstrates the political and relational possibilities of extraordinary affects, and the importance of the stories that evoke them.

The final chapter, *Inviting* serves as an epilogue to this collection. Here, we offer some reflections as an encouragement to the reader, to practice storying with place differently, collectively and relationally. And, to carefully and responsibly story what matters.

References

Ahmed, S. (2006). *Queer phenomenology: orientations, objects, others*. Duke University Press.

Banivanua Mar, T. (2012). Settler-colonial landscapes and narratives of possession. *Arena Journal, 37*(38), 176–198.
Bawaka Country, Burarrwanga, L., Ganambarr, R., Ganambarr-Stubbs, M., Ganambarr, B., Maymuru, D., Lloyd, K., Wright, S., Suchet-Pearson, S., & Daley, L. (2022). Songspirals bring country into existence: singing more-than-human and relational creativity. *Qualitative Inquiry, 28*(5), 435–447.
Bawaka Country, Suchet-Pearson, S., Wright, S., Lloyd, K., & Burarrwanga, L. (2013). Caring as country: towards an ontology of co-becoming in natural resource management. *Asia Pacific Viewpoint, 54*(2), 185–197.
Behrendt, L. (2006). Rights, responsibilities and indigenous disadvantage. Impact, (Summer), p. 5–7.
Benjamin, J. (2019). Remembered into place. In The Routledge Handbook of Memory and Place (pp. 214–222). Routledge.
Birch, T. (1992). *'Nothing has changed': the making and unmaking of Koori culture*. Meanjin. Last accessed in July 2025. https://meanjin.com.au/essays/nothing-has-changed-the-making-and-unmaking-of-koori-culture/
Bondi, L. (2023). Into the thick of it: troubling case studies and researching close to therapeutic practice. In K. Tudor & J. Wyatt (Eds.), *Qualitative research approaches for psychotherapy* (pp. 3–14). Routledge
Bönisch-Brednich, B., & Meyer, S. (2024). The migrant storyteller. In B. Bönisch-Brednich, A. Christou, S. Meyer, M. J. Karner, & A. J. Escher (Eds.), *Migrant narratives: storytelling as agency, belonging and community* (pp. 185–204). Routledge.
Bridger, B., Kalogeropoulou, G., Marinia, H., & Díaz-Vicedo, N. (2022). *On spirals*. Sage Online Publication. https://doi.org/10.1177/13505068211068611
Cameron, E. (2012). New geographies of story and storytelling. *Progress in Human Geography, 36*(5), 573–592.
Carlson, B., & Farrelly, T. (2023). *Monumental disruptions: Aboriginal people and colonial commemorations in so called Australia*. Aboriginal Studies Press.
Carrasco, W. (2025). Listening to silent geographies of Indigenous self-determination. *The Professional Geographer (online), 77*, 478080. https://doi.org/10.1080/00330124.2025.2478080
Carter, R. G. S. (2006). Of things said and unsaid: power, archival silences, and power in silence. *Archivaria, 61*, 215–233. Special Section on Archives, Space and Power. https://archivaria.ca/index.php/archivaria/article/view/12541
Case, E. (2021). That which feeds: sacred solidarities and roots of resistance. *Asia Pacific Viewpoint, 62*(2), 132–142.

Cresswell, T. (2014). *Place: an introduction*. John Wiley & Sons.
Cresswell, T. (2019). *Maxwell street: writing and thinking place*. University of Chicago Press.
Crews, C. (2019). Earthbound social movements and the Anthropocene. *Journal for the Study of Religion, Nature and Culture, 13*(3), 333–372.
Cronon, W. (1992). A place for stories: nature, history, narrative. *The Journal of American History, 78*(4), 1347–1376.
Cruikshank, J. (1999). *The social life of stories: narrative and knowledge in the Yukon Territory*. UBC Press.
Daniels, S., & Lorimer, H. (2012). Until the end of days: Narrating landscape and environment. Cultural Geographies, 19(1), pp. 3–9.
De Leeuw, S., & Hawkins, H. (2017). Critical geographies and geography's creative re/turn: poetics and practices for new disciplinary spaces. *Gender, Place & Culture, 24*(3), 303–324.
Delgado, R. (2013). 'Storytelling for oppositionists and others: A plea for narrative', in R. Delgado and J. Stefancic (eds), Critical Race Theory: The Cutting Edge, 3rd edn, Philadelphia: Temple University Press, pp 71– 80.
Edensor, T., Kalandides, A., & Kothari, U. (2020). *The Routledge handbook of place*. Routledge.
Freistein, K., Gadinger, F., & Unrau, C. (2022). It just feels right. Visuality and emotion norms in right-wing populist storytelling. *International Political Sociology, 16*(4), olac017. https://doi.org/10.1093/ips/olac017
Frenette, A. (2024). Story-listening as methodology: a feminist case for unheard stories. *Gender, Place & Culture, 31*(3), 379–399.
Gegeo, D., & Watson-Gegeo, K. (2001). 'How we know': Kwara'ae rural villagers doing Indigenous epistemology. *Contemporary Pacific, 13*(1), 55–88. https://www.jstor.org/stable/23718509
Gibson, R. (2010). The known world. *Text, 14*(Special 8), 1–11. https://doi.org/10.52086/001c.31508
Gomel, E. (2000). Aliens among us: fascism and narrativity. *Journal of Narrative Theory, 30*(1), 127–162. https://doi.org/10.1353/jnt.2011.0005
Götsch, B., & Palmberger, M. (2022). The nexus of anthropology and narrative: ethnographic encounters with storytelling practices. *Narrative Culture, 9*(1), 1–22.
Hamilton, J., Potter, E., & Quigley, K. (13 May 2024). Do stories need critics? Environmental storyism and the ends of ecocriticism.
Hancox, D. (2021). *The revolution in transmedia storytelling through place*. Routledge.
Haraway, D. J. (2016). *Staying with the trouble*. Duke University Press.

Harris, D. M. (2020). Telling stories about climate change. *The Professional Geographer, 72*(3), 309–316.
Harris, D. M. (2024). Stories, crisis, and meaning-making: storying possibility and community in the terrain of cultural struggle. *cultural geographies, 31*(1), 33–46.
Hassett, D. (2024). Rupture and reconciliation: the neoliberal logics of Emmanuel Macron's colonial memory policies. Modern & Contemporary France, 32(1), 31–53.
Hones, S. (2022). *Literary geography*. Routledge.
Houston, D. (2013). Environmental justice storytelling: angels and isotopes at Yucca Mountain. *Nevada. Antipode, 45*(2), 417–435.
Houston, D., & Vasudevan, P. (2017). Storytelling environmental justice: cultural studies approaches. In R. Holifield, J. Chakraborty, & G. Walker (Eds.), *The Routledge handbook of environmental justice* (pp. 241–251). Routledge.
Hughes, K., & Moscardo, G. (2024). Once upon a time: the impact of storytelling on connecting people to natural landscapes. *Environmental Education Research, 30*(2), 235–250. https://doi.org/10.1080/13504622.2023.2246695
Jackson, M. (2002). *The politics of storytelling: violence, transgression, and intersubjectivity*. Museum Tusculanum Press.
Kanngieser, A. M., Soares, E., Rubis, J., Sullivan, C. T., Graham, M., Williams, M., Palis, J., Tynan, L., Daley, L., Blacklock, F., Greenhough, B., Suchet-Pearson, S., Wright, S., Lloyd, K., & Marshall, U. B. (2024). Listening to place, practising relationality: embodying six emergent protocols for collaborative relational geographies. *Emotion, Space and Society, 50*, 101000. https://doi.org/10.1016/j.emospa.2024.101000
Keeshig-Tobias, L. (1990). The magic of others, In Libby Scheier, Sarah Sheard & Eleanor Wachtel, language in her eye: Views on writing and gender by canadian women writing in. Coach House Press (1990).
Kothari, U., & Klein, E. (2023). *Advanced introduction to critical global development*. Edward Elgar.
Latour, B. (2014). Agency at the time of the Anthropocene. *New Literary History, 45*(1), 1–18.
Liboiron, M. (2022, March 2). *Citational politics in tight places*. CLEAR. Last accessed in July 2025. https://civiclaboratory.nl/2022/03/02/citational-politics-in-tight-places/

Lorimer, H. (2003). Telling small stories: spaces of knowledge and the practice of geography. *Transactions of the Institute of British Geographers, 28*(2), 197–217.

Massey, D. (2013). *Space, place and gender*. John Wiley & Sons.

Masud, N. (2023). *A flat place: A memoir.* London: Penguin

McQuire, A. (2024). *Black Witness: the power of Indigenous media.* University of Queensland Press.

Meretoja, H., & Davis, C. (2017). *Storytelling and ethics*. Routledge.

Miles, J. (2019). Historical silences and the enduring power of counter storytelling. *Curriculum Inquiry, 49*(3), 253–259. https://doi.org/10.1080/03626784.2019.1633735

Morgan, C. (2019). Walking, writing, reading place and memory. In S. De Nardi, H. Orange, S. High, & E. Koskinen-Koivisto (Eds.), *The Routledge handbook of memory and place* (pp. 261–267). Routledge.

Moss, M., & Thomas, D. (2021). Theorising the silences, in archival silences: missing, lost and, uncreated archives. In M. Moss & D. Thomas (Eds.), *Archival silences: missing, lost and, uncreated archives* (pp. 10–25). Routledge.

Mott, C., & Cockayne, D. (2017). Citation matters: mobilizing the politics of citation toward a practice of 'conscientious engagement'. *Gender, Place & Culture, 24*(7), 954–973.

Okri, B. (2015). Why do we tell children stories? *Africa Writes.* Last accessed in July 2025. https://africawrites.org/blog/ben-okri-why-do-we-tell-children-stories/

Okri, B. (2015b, November 4). Under the sun: a meditation by Ben Okri on stories." *The Irish Times.* Last accessed in July 2025. https://irishtimes.com/culture/books/under-the-sun-a-meditation-by-ben-okri-on-stories-1.2416769

Ortiz, C. (2023). Storytelling otherwise: decolonising storytelling in planning. *Planning Theory, 22*(2), 177–200. https://doi.org/10.1177/14730952221115875

Perec, G. (1997). *Species of spaces and other pieces.* Penguin.

Rose, M. (2016). Envisioning the future: Ontology, time and the politics of non-representation. In Taking-place: Non-representational theories and geography (pp. 341–361). Routledge.

Rubis, J. M., & Theriault, N. (2020). Concealing protocols: conservation, Indigenous survivance, and the dilemmas of visibility. *Social & Cultural Geography, 21*(7), 962–984.

Sandercock, L. (2003). Out of the closet: the importance of stories and storytelling in planning practice. *Planning Theory & Practice, 4*(1), 11–28. https://doi.org/10.1080/1464935032000057209

Smith, A. M. (2014). Old fieldwork, new ethnography: taking the stories out of the bag. *Qualitative Inquiry, 20*(5), 699–708. https://doi.org/10.1177/1077800413513729

Smith, T., & Fiore, J. (2010). Landscape as narrative, narrative as landscape. *Studies in American Indian Literatures, 22*(4), 58–80.

Smith, A. S., Smith, N., Daley, L., Wright, S., & Hodge, P. (2021). Creation, destruction, and COVID: Heeding the call of country, bringing things into balance. Geographical Research, 59(2), 160–168.

Somerville, W., McKenzie, V., Fuller, L., Godden, N. J., Harrison, A., Isaacs-Guthridge, R., & Turner, B. (2023). "Kind regards": negotiating connection to Country and place through collective storying. *Australian Journal of Environmental Education, 39*(3), 336–348.

Spearim, B. (Host). (2025). Awesome black. *Frontier War Stories Series*. [Audio Podcast]. Last accessed in July 2025. https://awesomeblack.org/series/frontier-war-stories/

Springgay, S., & Truman, S. E. (2018). On the need for methods beyond proceduralism: Speculative middles, (in) tensions, and response-ability in research. *Qualitative Inquiry, 24*(3), pp. 203–214.

Stadler, J., Mitchell, P., & Carleton, S. (2016). *Imagined landscapes: geovisualizing Australian spatial narratives*. Indiana University Press.

Storying Geography Collective (Wright, S., Palis, J., Osborne, N., Miller, F., Kothari, U., Henrique, K.P., Everingham, P., Borovnik, M.) (2023). Storying pandemia collectively: sharing plural experiences of interruption, dislocation, care, and connection, GeoHumanities, 9(1), 1–23.

Teaiwa, K. (2015). Ruining Pacific Islands: Australia's phosphate imperialism. *Australian Historical Studies, 46*(3), 374–391. https://doi.org/10.1080/1031416X.2015.1082609

Todd, Z. (2016). An indigenous feminist's take on the ontological turn: "Ontology" is just another word for colonialism. *Journal of Historical Sociology, 29*(1), 4–22.

Trouillot, M. R. (1995). *Silencing the past*. Beacon Press.

Tuan, Y. F. (1975). Place: an experiential perspective. *Geographical Review, 65*(2), 151–165.

Tynan, L., & Bishop, M. (2023). Decolonizing the literature review: a relational approach. *Qualitative Inquiry, 29*(3-4), 498–508.

Van Dooren, T., & Bird Rose, D. (2016). Lively ethography: storying animist worlds. *Environmental Humanities, 8*, 77–94.

Vannini, P. (2015). Non-representational ethnography: New ways of animating lifeworlds. *Cultural Geographies, 22*(2), pp. 317–327.

Vasudevan, P., Ramírez, M. M., Mendoza, Y. G., & Daigle, M. (2023). Storytelling earth and body. *Annals of the American Association of Geographers, 113*(7), 1728–1744.

Watego, C. (2021). *Another day in the colony*. University of Queensland Press.

Watson, N. (2024). Counter-storytelling as critical praxis. In D. Bargallie & N. Fernando (Eds.), *Critical racial and decolonial literacies: breaking the silence* (pp. 190–202). Bristol University Press.

Wilcox, H., McWatters, K., & Thompson, A. (1990). *The body and the text: Helene Cixous: reading and teaching*. Harvester Wheatsheaf.

Wright, A. (2016, March 15). *What happens when you tell somebody else's story?* Meanjin. https://meanjin.com.au/essays/what-happens-when-you-tell-somebody-elses-story/

Wright, S. (2022). Poetry as decolonial praxis. In K. Sims, N. Banks, S. Engel, P. Hodge, J. Makuwira, N. Nakamura, J. Rigg, A. Salamanca, & P. Yeophantong (Eds.), *The Routledge handbook of global development* (pp. 713–723). Routledge.

Wright, S., Lloyd, K., Suchet-Pearson, S., Burarrwanga, L., Tofa, M., & Country, B. (2012). Telling stories in, through and with country: engaging with Indigenous and more-than-human methodologies at Bawaka, NE Australia. *Journal of Cultural Geography, 29*(1), 39–60.

Wright, S., Suchet-Pearson, S., Lloyd, K., Burarrwanga, L. L., & Burarrwanga, D. (2009). 'That means the fish are fat': sharing experiences of animals through Indigenous-owned tourism. *Current Issues in Tourism, 12*(5-6), 505–527. https://doi.org/10.1080/13683500903042907

Yandaarra with Gumbaynggirr Country (Smith, A.S., Marshall, U.B., Smith, N., Wright, S., Daley, L. & Hodge, P.) (2022). Ethics and consent in more-than-human research: some considerations from/with/as Gumbaynggirr Country, Australia. Transactions of the Institute of British Geographers, 47(3), 709–724.

Youatt, R. (2017). Personhood and the rights of nature: the new subject of contemporary earth politics. *International Political Sociology, 11*(1), 39–54.

2

Responding: Absence, Loss and Power

Fiona Miller

Abstract Stories evoke complex emotional geographies of place that connect across different scales, from the intimate to the global. Stories enable us to draw connections between major forces of injustice, such as climate change, and uneven consequences for particular people and places, collapsing the distance seemingly stretched across space and time between contribution and consequence. Through stories, attention can be focused upon processes that contribute to harm, and potentially marginalised experiences and acts of resistance and repair, invoking a politics of responsibility. In this chapter I consider how stories of absence and loss reveal what people value, and how such stories can act as a guide towards other ways of being in the world that challenge dominant power relations. The chapter argues that through engagement with such stories, we can appreciate the power of stories to counter hegemonic power and enact alternative imaginaries and relations with place.

Keywords Politics · Imaginaries · Emotional geographies · Resistance · Absence and loss

U. Kothari et al., *Stories of Place*, https://doi.org/10.1007/978-981-95-4921-4_2

Introduction

In this chapter, I focus on how relations with place are imbued with emotions, memories, knowledges, hopes and imaginaries. Stories and storytelling reveal the shifting emotional geographies of place—what we value, imagine, long for, notice, overlook, remember and forget. Stories can capture the pluralistic nature of meaning and significance of place in the everyday (Storying Geography Collective et al., 2023). Seemingly small stories can come together and connect with bigger stories. Through story, people and their grounded experiences can evoke experiences of loss, of migration, of identity, of belonging and exclusion, of safety and fear, enabling emotional connections to particular places to resonate across time and space.

The stories we tell and the stories we carry demonstrate intimate connections between place and ourselves – we form attachments to, learn from and are affected by place. We are implicated in and witness damage to place and are harmed by particular places. We also have responsibilities to care for and protect place. Stories not only describe place and enable us to learn and feel but also create and change places through imaginaries. Stories 'make worlds' as they evoke imaginaries of place and create relations of care—for place, each other and the living world. Counter imaginaries of place reveal the ways power is enacted through storying, demonstrating the political role stories play in asserting and resisting dominant framings, interests, histories, paradigms and ideologies.

Stories are also a way of responding to change. They help us make sense of bewildering, unjust, uncanny and unprecedented changes reconfiguring relations between people, place and the living world. Stories can act as maps to help us navigate this change. To recollect a path taken, to decide future paths and to chart new paths. They can help us get lost, if we so wish, to escape from the ordinary and painful, holding out the possibility of being reorientated somehow, sometime.

Through consideration of the emotional geographies of place I want to draw attention to how stories and storytelling are a way of responding to absence and loss and can be a form of care - care that can be enacted collectively through sharing, reading, listening and interpreting. Through sharing stories, for instance, in research and teaching contexts, relations of care can be foregrounded. The trust people imbue in the power of storytelling, notably in the context of story-based research methods, is a way of responding to changing relations with place. Care can be enacted through the acknowledgement of the emotional affect of absence and loss, as well as identifying and responding to people's well-being needs. In this chapter, through a focus on harm to and loss of places due to climate change, I consider the role of stories in helping understand these processes as well as responses to such loss calling for a widening of our gaze from the sites of harm to the multiscale processes that shape relations with place. Through the themes of noticing, witnessing, meaning-making, responsibility and imagining I highlight the different ways stories capture our responses to these climate change times.

Noticing

Stories have the capacity to reveal patterns across space, making manifest the changing nature of social, cultural, economic and political relations with place. Stories can transcend apparent boundaries between the intimate and corporeal to larger, even global, scales (Vasudevan et al., 2023). They allow us to time travel to distant pasts and futures bringing relations with the past and future within our grasp and ken, opening up possibilities to notice and to read the text of places in our present lives in new and potentially consequential ways. Stories have the capacity to collapse the gulf of distance between far-off and proximate places, such that recognition of what is common and different generates constructive comparisons and insights. Yet, stories of place also have the capacity to alienate, inflame suspicion and fear, and provoke feelings of resentment and distrust, depending on the stories told, the motivation of the storyteller and the reception of the audience.

Stories of place highlight the things people value, and their observations about how a place changes with the seasons and over time. In the noticing of seasonal rhythms, the quotidian practices of people and non-humans, the constant shifting of sound, light and shadow, and the disruptions or ruptures to such patterns it is possible to gain a deeper sense of place. Yet, to notice one needs to slow down and to have regard and an openness to the stories place offers.

Stories can detail the embodied and emotional implications of temporal and spatial changes. The noticing of such connections can generate new understandings and politics, as captured in my story. The hours I have spent cycling, focusing on my breathing, my pedal strokes and the changing terrain around me, gave me the space to think and craft the following story. I wrote it after an intense period of research where I had listened to people's stories of their experiences with severe floods. It is an attempt to make sense of the emotional affect of the changes to place that I have experienced and witnessed.

It is the absences that have come to mark these climate-changed times.
We are becoming familiar with the stories of the bogong moths no longer swarming, the clean windscreens after long country drives, the Christmases without colourful beetles, and the summers without fear of fires. These absences slowly creep up on you. Extinctions, though now anticipated, are rarely announced. Yet, absences still leave traces. These traces may be subtle, in the form of memories and feelings, whereas some absences manifest as voids.
Extractivism underpins this brief time in human history as the fossils, that took millions of years to form, have been rapidly extracted and combusted in the space of just a few generations to fuel particular economies and lifestyles, at the expense of others. This extractivism is predicated on exclusion and dispossession. Yet, once value is extracted from a landscape, voids form – these voids loom large now where forests once covered the soils layered over coal. These voids have come to redefine these once lively places. The uneven, ancient topography of undulating hills and the river-formed valleys are now pocked by voids. Light and shadow falls differently, water flows and pools in unusual ways, and new and old plants and animals compete to find novel niches. Very occasionally there have been attempts to fill these voids, to terraform the hazards of deep holes away – to construct an artifice of what once was. Yet, somehow, the accumulation and concentration of wealth that has accompanied the expansion of these voids is inaccessible when the scheduled time for site remediation arrives. After the rumble of the bulldozers, the graders and drilling machinery ends there is the silence of the void. The communities that remain now struggle to repair these harmed places and to transition on to the next thing, still yet undefined.

> *Yet, there are other voids which mark this epoch. These are smaller and more dispersed and mundane than those associated with mining, but somehow are connected with the extraction and combustion of fossil fuels through a complex circuitous chain of relations.*

I ride my bike. I ride it a lot. As I ride I ponder things. I notice the landscape. I keep my eyes on the road and my ears open for angry or distracted motorists. The intense la niña event in eastern Australia in 2022 inconvenienced my bike riding. It rained. It rained a lot and it rained heavily. We thought, as with the fires and the smoke of the 2019-2020 Black Summer, that the rain would never end. There was mould. There was moss. There were floods. Terrible floods. People in harm's way were … harmed. As one system collapses, other systems fall apart. One such system that fell apart was the Sydney rail system as tracks and wires were washed away or damaged by the rains. I, like so many others, had previously relied upon this system to transport me from place to place. The delays and disruptions became too much, so I decided it was preferable to ride my bike rather than to wait for a train that might not come or might not arrive at its destination.

> *Yet with the rain, came the potholes. Little voids opened up. These potholes, though unremarkable and mundane, are hazardous to cyclists, causing injury, punctures, buckles, and accidents. In the heavy rain of 2022, the chat for my cycling club lit up with warnings of which roads to avoid due to potholes and which numbers to call to report holes to be repaired. Yet, many of these voids remain unfilled months later. The investment in public roads for private cars, that got us into this climate mess, does not quite stretch to maintaining these roads through the climate crisis.*
>
> *Potholes are not just an issue for woke, city greenies in Lycra; the absence of sufficient public funds to support post-flood recovery has meant severely flood damaged roads and bridges remain a major concern for remote and regional communities affected by the eastern Australian flood disaster of 2022–2023. The delay in repair works has intensified the isolation and increased the travel times and cost for many rural communities. Local councils have not adequately budgeted for the unfolding climate crisis. Again, somehow the funding to fill the voids is absent. Communities have variously stepped into the void of inadequate formal recovery, working to repair less obvious damage*

associated with the accumulated mental, financial and emotional toll of the floods.

It is these absences that I have come to notice. As I ride my bike, from place to place, I often ponder: does the unremarkable pothole signify a tipping point? Do these voids mark not just a point in the landscape but a point in time where our social capacity for recovery and repair can no longer be assured to keep pace with the frequency of extreme weather events? Alternatively, might our growing awareness and experience of loss open up new connections and political possibilities for repair? (Fig. 2.1)

This story evokes complex emotional geographies of place that connect across different scales, from the intimate to the global, and across space and time; from the scars left on communities and landscapes by coal mining to the mundane and barely noticed pothole to the complex political economies of essential repair and maintenance amidst the unfolding climate crisis. The story is an attempt to bear witness, to show the connections between the harm set in train by extractivist economies and the distant and varied implications for people's particular

Fig. 2.1 Pothole and car on bike lane. *Source* F. Miller

relations with place. The story emerged from the possibility to experience place differently and the chance to slow down that was afforded to me through cycling. Cycling has been one of my ways of coping with the emotional impacts of the climate breakdown we are all living through. As Anna Tsing writes,

Without stories of progress, the world has become a terrifying place. The ruin glares at us with the horror of its abandonment. It's not easy to know how to make a life, much less avert planetary destruction. Luckily there is still company, human and not human. We can still explore the overgrown verges of our blasted landscapes - the edges of capitalist discipline, scalability, and abandoned resource plantations. We can still catch the scent of the latent commons - and the elusive autumn aroma (2015, p. 253).

It is in this art of noticing that we can find lively stories of place beyond the voids, the damage and the ruins. As Cameron (2012, p. 578) writes, it is these particularities that "reveal geographies that are simply not apparent when our attention is trained on the institutional, the epistemic, and the discursive". Stories give us the opportunity to notice, think about and imagine other relations with place.

Witnessing

Stories are a powerful way to bear witness to the harm wrought by processes of injustice and their uneven expression across space and time. As expressed by van Dooren and Bird Rose, "Ethographic storytelling is about the arts of becoming-witness, which include both attention to others and expression of that experience: to stand *as* witness and actively to *bear* witness. As we are seized, so we bear witness in order that others may be seized, telling stories that draw audiences into others' lives in new and consequential ways, stories that cultivate the capacity for *response*" (2016, p.89). Here, van Dooren and Bird Rose express the hope that stories are consequential, drawing others in to respond in appropriate ways.

This dedication to bearing witness is one way of responding to the unjust decisions, actions, values and economies transforming people and places. Van Dooren and Bird Rose further note that.

Storytelling is one of the great arts of witness, and in these difficult times telling lively stories is a deeply committed project, one of engaging with the multitudes of others in their noisy, fleshy living and dying. It is the aim of lively ethographies to seize our relational imagination. It is an engagement with the joys, passions, desires, and commitments of Earth others, celebrating their ethea in all their extravagant diversity (2016, p. 91).

Through story, the act of witnessing can generate political possibilities by evoking emotional reactions that, potentially, lead to action. The political act of witnessing through story is forcefully expressed in the book *Black Witness* (McGuire, 2024a) written by Aboriginal journalist Amy McGuire, which exposes the ongoing violence and racism experienced by Indigenous people in so-called Australia. McGuire documents, through a series of essays, why the Black Witness should be believed. In an extract from the book, McGuire writes,

In the absence of our stories, we were spoken over, and for. But our ancestors were witnesses, and they remained on this country. They tell their stories through those who lived, those who survived. As Black Witnesses, we embody that history: we are a conduit for our ancestors. The crimes of a country are told and retold upon the waters, kept in the mountains, retained in the soil of our land. We are sovereign because we belong to this land and live relationally with it (2024b, para. 3).

Stories as such can be the catalyst that results in the formation of new political constellations for justice. In another setting, as Harris (2022) reminds us, Rachel Carson's powerful use of story (together with science) in *The Silent Spring* (Carson, 1962) evocatively captured the mass death of birds due to toxic chemical use in industrial agriculture. Carson's storytelling was hugely consequential for the conservation movement in the United States and beyond, leading to improvements in subsequent environmental policies and regulations.

Yet, in our intensely storied world, the question as to why such witnessing does not evoke more change towards careful and reparative relations with place and the living world is perplexing. Indeed, we are witnessing a deliberate turning away from stories of harm and loss—it is just too much for some, especially young people, to bear. Perhaps stories of love, joy, survival and care are what we need to be more attuned to. As Eve Tuck (2009) has written, the deliberate framing of some places and people as vulnerable, and the constant focus on damage, can further compound that damage. Such stories can pathologise particular people and places, even when, inadvertently, damage focused research seeks to draw attention to the historical, colonial and political-economic causes of that damage. Instead, Tuck suggests an "alternative to damage-centered research is to craft our research to capture desire instead of damage. I submit that a desire-based framework is an antidote to damage-centered research (2009, p. 416)". As such, we need to hold space for stories of harm and loss, when these are the stories people wish to tell, but ensure such stories do not foreclose the possibility to simultaneously engage with stories of place that capture diverse emotions like joy, desire, repair, care and survival.

Meaning-Making

Diverse emotions underpin the stories people share of the places, and the living things, they are entangled with. While love, sadness and nostalgia are ever present, anger is also often expressed, particularly when there is a sense of shock and injustice. In the context of disasters, including those induced by the global climate emergency, beyond the stories of spectacle and heroism that tend to dominate popular accounts, there are the quiet and careful stories of reflection that capture people's attempts to make meaning from the disorientation that comes in the aftermath.

In such stories, there is a grappling with explanations of the apparently random and sometimes violent nature of the rupture in people's lives. There is an attempt to make some sense from seemingly pointless destruction, by piecing together threads of meaning from events, signs, actions, objects and relationships. Drawing on Brockman (2013), Harris writes, "Stories are mechanisms to help us deal with uncertainty — narratives told through drawing, song, or ritual have long been used to reduce fear, allay doubt, and pose and answer questions about the unknown" (2022, p. 837). As Vasudevan et al. write,

> Storytelling is how people make sense of the world and our place in it, an iterative process of interpreting reality through observation and the exchange of ideas. Storytelling, as a means of theorizing the structural from embodied experience, is a fundamental methodological intervention of Indigenous, Black, and third world and women of color feminist traditions (2023, p. 1729).

Underpinning the sharing of stories of climate disasters is a hope that sense can be made of the mess and that lessons can be drawn to ensure others might avoid similar experiences. This faith in storytelling reflects people's willingness to relive their pain in order that their stories may help others. In research contexts, it reflects a hope that something good can come from their experience—an effort to ensure their experience is not ignored, overshadowed or forgotten, subject to a "public amnesia and collective forgetting" (Houston, 2013, p. 426). In trauma-informed research, gentle story-based methods can provide a safe and non-judgemental way for people to express what is meaningful for them and to make sense of their changing relations with place—their pain and their joy. There are countless stories expressed in the aftermath of disasters, reflecting people's diverse material circumstances, emotional responses and environmental encounters. As people try to make sense of their experiences, it is quite common though for people to focus on the generosity of strangers, others less fortunate than themselves, the importance of family and community connections, frustrations with bureaucratic responses, their attachment to particular (though not necessarily valuable) objects, and the missed warning signs, such as the odd behaviour of animals. Sometimes, amidst the confusion, anger and disorientation that come with extreme events, when the familiar characteristics of place are altered, stories can reveal people's vulnerability to conspiracy narratives, harmful ideologies and the politics of division, especially when institutions fail and people's sense of security slips away. As hard as it is to hear such stories, we have a responsibility to nonetheless bear witness, as they capture the unravelling of trust and certainty and are key to thinking through appropriate responses that address the causes of harm. Yet, as Tuck (2009) and Tuck and Yang

(2014) have warned, care is required to ensure stories of damage are not the only ones we wish to hear and do not affix to place and communities in harmful ways. Drawing on bell hooks' (1990, p. 343) critique of the fixation, extraction and possession of the stories of pain of marginalised people in social science research, Tuck and Yang write,

Yet the forces that invite those on the margins to speak also say, 'Do not speak in a voice of resistance. Only speak from that space in the margin that is a sign of deprivation, a wound, an unfulfilled longing. Only speak your pain' (2014, p. 228).

Thus, there is a responsibility to be open to the stories people wish to share in their efforts to make sense of climate crises, to not predetermine or foreclose these. As Solnit reminds us, how we remember disasters is important for how we prepare for future disasters, "so getting the story right matters for survival as well as for justice and history" (quoted in Houston, 2016, p. 446). In so doing, we need to find ways to widen our gaze to capture a politics of responsibility.

Responsibility

Rather than fixating on places of harm, stories enable both the teller and listener to draw connections between major forces of injustice, such as climate change, and their consequences for particular people and places, collapsing the distance seemingly stretched across space and time between contribution and consequence. The narrative arc is extended, the aperture widened, to take in historical and structural processes. As novelist Amitav Ghosh has written, climate change represents "forces of unthinkable magnitude that create unbearably intimate connections over vast gaps in time and space" (Ghosh, 2016, p. 63). Stories thus enable us to address the challenges of capturing the nature of such slow violence. As Rob Nixon writes,

The explicitly temporal emphasis of slow violence allows us to keep front and centre the representational challenges and imaginative dilemmas posed not just by imperceptible violence but by imperceptible change whereby violence is decoupled from its original causes by the workings of time (Nixon, 2011, p. 11).

It is in the noticing, witnessing and listening to people's stories, and drawing connections across space and time, that it is possible to consider the politics of responsibility for the defining stories of our current epoch—of loss, of displacement, of repair, of hope and of care. It is particular people, particular practices and particular economic and political decisions that shape geographies of harm and loss. These are central themes in much of the work of contemporary human geographers who engage with story as method and stories and storytelling as material. As Hamilton et al. (2024, p. 2) write, "story's multifarious rise to prominence amidst the environmental humanities and social sciences has represented a pivotal and often galvanising response to perceived … shortcomings in standard modalities of apprehension, interpretation, and criticism". Stories certainly complement traditional methods of analysis in human geography, but more than this, they also access and convey embodied knowledge and the emotional dimensions of relations with place. Stories enable us to do geography differently and to understand and address the politics of responsibility. Harris argues that "story can help to uncover the complex topographies of power, subjective and relational dimensions, as well as inequalities and emotions that are central to everyday and material struggles over resource use, access, and conditions" (Harris 2022, p. 848). She suggests that in addition to capturing what changes are occurring, a "focus on narrative, story, and storytelling usefully helps to shift analytics" (ibid) towards the meaning people attach to such changes, and emerging themes. They provide an opportunity to imagine otherwise, capturing people's aspirations for different kinds of futures, and potentially guiding us towards other ways of being in the world, towards other ethics of care and relations to place.

Imagining Otherwise

Storytelling is seen by Vasudevan et al. (2023), as "a fundamental practice of imagining the world otherwise. To survive and rebuild as 'we drift as a species toward an unparalleled catastrophe' (Wynter & McKittrick, 2015, p. 18), we must rethink our place in the world and how we envision the societies we collectively build" (p. 1738). Stories of memories and aspirations associated with place can evoke something that is imaginary—a memory of what once was or a longing for something to come. This 'world making' potential of stories can contribute to evoking different kinds of relations with place and each other, and has been seized upon by many scholars concerned with the injustice of contemporary relations with place (see Harris, 2022; Houston, 2013, 2016; Vasudevan et al., 2023).

Recognising that some stories matter more than others, and that some stories of place overshadow or silence others, highlights the political aspect of storytelling. As Houston writes, stories are "full of resonance, memory and evidence, they reflect discernible and powerful truths" (Houston, 2013, p. 433). As political geographers, engaging with 'counter' imaginaries is thus crucial to challenging dominant stories and relations with place. In a time of climate emergency, where denialism is rapidly being replaced by defeatism, stories have world making potential and enable us to imagine and act otherwise.

Stories can orientate us away from damage to joy. Can we imagine a world where the joy of cycling fuels the desire to eschew the convenience of cars that transforms our cities and economies from ones reliant on extraction and combustion to ones based on care and repair? Stories can evoke a sense of responsibility and care for place, but can also simultaneously highlight the politics, economics and ethics of extraction that result in others evading their responsibility for harm to place. In conclusion, it is fitting to acknowledge that stories continue to unfold and help guide us through difficult times. Referring to writer Ursula Le Guin's *The Carrier Bag Theory of Fiction,* Anna Tsing writes that "storytelling might pick up diverse things of meaning and value and gather them together, like a forager rather than a hunter waiting for the big kill. In this kind of storytelling, stories should never end, but rather lead to further stories" (Tsing, 2015, p. 256).

References

Brockman, R. (2013). Only stories matter: The psychology and neurobiology of story. *American Imago, 70*(3), 445–460.

Carson, R. (1962). *Silent spring*. Houghton Mifflin.

Cameron, E. (2012). New geographies of storytelling. *Progress in Human Geography, 36*(5), 573–592.

Storying Geography Collective (Wright, S., Palis, J., Osborne, N., Miller, F., Kothari, U., Henrique, K.P., Everingham, P. & Borovnik, M.) (2023). Storying pandemia collectively: sharing plural experiences of interruption, dislocation, care, and connection. *GeoHumanities*, 9(1), 1–23.

Ghosh, A. (2016). *The great derangement: climate change and the unthinkable*. University of Chicago Press.

Hamilton, J., Potter, E., & Quigley, K. (2024). Do stories need critics? Environmental storyism and the ends of ecocriticism. *Textual Practice, 38*, 1–23. https://doi.org/10.1080/0950236X.2024.2348066

Harris, L. M. (2022). Towards enriched narrative political ecologies. *Environment and Planning E: Nature and Space, 5*(2), 835–860.

Houston, D. (2013). Environmental justice storytelling: angels and isotopes at Yucca Mountain. *Nevada. Antipode, 45*(2), 417–435.

Houston, D. (2016). Crisis is where we live: environmental justice for the Anthropocene. In J. Goodman & J. Marshall (Eds.), *Crisis, movement, management: globalising dynamics* (pp. 97–108). Routledge.

McGuire, A. (2024a). *Black witness: the power of indigenous media*. University of Queensland Press.

McGuire, A. (2024b). *Extract: black witness*. The Australia Institute. https://australiainstitute.org.au/post/extract-black-witness-by-amy-mcquire/. Access Date: 13 July 2025.

Nixon, R. (2011). *Slow violence and the environmentalism of the poor*. Harvard University Press.

Tsing, A. L. (2015). *The mushroom at the end of the world: on the possibility of life in capitalist ruins*. Princeton University Press.

Tuck, E. (2009). Suspending damage: a letter to communities. *Harvard Educational Review, 79*(3), 409–428.

Tuck, E., & Yang, K. W. (2014). R-words: refusing research. In P. Django & M. T. Winn (Eds.), *Humanizing research: decolonizing qualitative inquiry with youth and communities* (pp. 223–247). Sage.

Van Dooren, T., & Bird Rose, D. (2016). Lively ethography: storying animist worlds. *Environmental Humanities, 8*(1), 77–94.

Vasudevan, P., Ramírez, M. M., Mendoza, Y. G., & Daigle, M. (2023). Storytelling earth and body. *Annals of the American Association of Geographers, 113*(7), 1728–1744.

Wynter, S., & McKittrick, K. (2015). Unparalleled catastrophe for our species? Or, to give humanness a different future: Conversations. In Sylvia Wynter: On being human as praxis, ed. K. McKittrick, 9–89. Durham, NC: Duke University Press.

3

Imagining: Disrupting Colonial Imaginaries, Creating Postcolonial Stories

Uma Kothari

Abstract Stories emerge from our own imaginations and are influenced by the imaginaries of others. This chapter reveals how dominant and authoritative colonialist stories created in particular cultural, political and historical contexts can be displaced and challenged over time. Drawing on stories of an extraordinary road trip from Britain to India that my parents undertook in 1955, this chapter shows how colonial narratives, in which they had been enmeshed as colonised subjects, were entrenched in fantastical imaginaries of Britain as a modern, civilised and even glorious world. It was these accounts that fed their aspirations to travel from the colonial periphery to the imperial centre and moulded their expectations of what they would see and experience. However, as they encountered Britain up close, these preconceptions were disavowed, gradually being replaced as they created their own postcolonial, counter-stories.

Keywords colonialism · imaginaries · postcolonial stories · landscapes · English countryside

U. Kothari et al., *Stories of Place*, https://doi.org/10.1007/978-981-95-4921-4_3

As a child I mistook my parents' repeatedly thwarted search for the English countryside as evidence of their unworldliness.

In 1960s London, summer Sundays were characterised by long family car journeys in search of an idyllic rural spot for a picnic.

Having been educated in India during British rule, my parents had imbibed powerful colonial romanticisms that gave rise to a desire to visit the English countryside - always imagined as a singular, mythical realm, the English countryside. As the novelist Hanif Kureishi (2014) acknowledges, for many Indian migrants, the English countryside was a foreign land; they were ensnared in a complex relationship in which it seemed profoundly strange, even frightening, yet also intensely alluring.

And so, on these Sundays we drove away from the capital in search of a magical landscape that would epitomise green and pleasant England.

Squashed into our two-door, red Morris Mini Minor car, equipped with picnic hamper and thermos flask, and with us three sisters in the back, we peered out of the vehicle's tiny windows as London's dense urban environment turned into suburban sprawl, then into an extensive succession of semi-rural commuter villages before settling into more expansive green stretches of fields and trees.

And we indefatigably drove out and around country lanes, searching, looking, and usually deciding against stopping.

A rural idyll never quite materialised during these Sunday motoring meanders; after driving for hours we would return to our home in North London, hot, bickering and frustrated as once more, the sublime English countryside had eluded us. The grass was too rough, the hills did not roll, the sun rarely shone for long and if we did venture out of the car, much time was spent fending off the hordes of ants attracted by the delicious smells emanating from our tiffin boxes. But, as we spread a rug on the grass in our suburban garden and tucked into roti and curry, it no longer mattered – and there was always next Sunday.

We recurrently embarked on these car journeys, but each time with declining enthusiasm and the growing tinge of anticipated disappointment as the quest for an idealised, flawless English countryside repeatedly ended in defeat and disillusionment. That is until, after numerous failed attempts, my parents and their friends found the perfect place, only a few miles from where we lived.

The Ruislip Lido.

This bucolic retreat, situated within the London Borough of Hillingdon, possesses a lake and artificial beach. Originally opened as a feeder reservoir for the Grand Junction Canal in 1811, the site was reopened in 1933 as a lido, offering facilities for swimming. We piled into the rowing boats available for hire from boathouses on the western shore to float upon the serene waters and took rides on a fairy-tale miniature steam railway that rattled through ancient woodland as it circled the lake.

My parents had found their English countryside; a curated, manicured place at the end of the Metropolitan underground line that felt safe, contained and ordered. We stopped searching and spent many long summer days - although photos depicting us with coats and scarves reveal that we might often have felt rather chilly - content that we had found a place that was English enough. We epitomised journalist Amit Roy's (2006) apt comment that 'Indians bring their own distinctive sensibility to what inspires them about the English countryside'. And so it was that my parents developed their own way of perceiving, desiring and making Ruislip Lido a familiar realm in which we could pleasantly linger on a summer's day.

I suspect, however, that they had simply transfigured the lido into our ideal rural location as our unrequited pursuit became incessant. But as far as I was concerned, our quest for the English countryside remained unfulfilled. The lido was fun but when I was invited for a day out motoring with my white British friends, their parents always seemed to find the glorious countryside that my parents were unable to locate; verdant meadows, sun-dappled woodland and meandering streams. It appeared as if they knew where to discover such places, but perhaps I simply imagined those sites at which we frolicked and picnicked as being more perfectly idyllic.

These contrasting adult competencies evoked in me rather ungenerous feelings about my parents. I did not appreciate that they and their friends had migrated far from their homeland, separated from their families and everything to which they were accustomed. They were still struggling to adjust to

> *a completely new country, with its unfamiliar conventions, tastes and leisure pursuits.*
>
> *Instead, I considered them unworldly. Untravelled, unadventurous and parochial.*

What these cautious, unambitious outings, and the safe choice of Ruislip as a weekend destination, could not reveal was that they had at one time been remarkably adventurous; bold, carefree and intrepid.

In May 1955, in their mid-20s and not long married, my parents travelled from Mumbai, India to Southampton, England on a passenger steamer. They boarded the boat-train from the port and on their arrival in London, purchased a second-hand Morris Minor car and spent the next six months motoring overland back to India. They had embarked on what they envisaged would be an adventure of a lifetime (Fig. 3.1).

While they had extraordinary stories to tell of their adventures on this extended road trip, for many years, their perceptions, experiences and reflections remained largely untold and unheard. They did not share their stories fully until later and I was not ready to hear them until later still.

In the late 1950s, a few years after they had returned to India following their travels, my father was offered a job in London, and they migrated to Britain. The teller of stories needs to be ready and willing to recount their tales, but there was little time and few circumstances for their nostalgic indulgences to be aired as they struggled with settling in a foreign country. So, for a while, they stowed away their memories of a more carefree, youthful phase in their lives.

And, during my primary school years when my parents would occasionally recall a memory of an encounter or incident that had taken place on that journey, I did not pay much attention, unable to imagine my parents as ever having been young and adventurous. I appreciate now that you can only hear stories when you are ready to listen, and I regret not being attentive in listening more carefully to the stories they imparted when I was younger. I was unable, as Larissa Behrendt (2021, p. 249) writes, to be able to listen deeply, "with respect - to hear, think, don't hurry, listen to the story – and consider what's in the spaces".

Fig. 3.1 My parents on their road trip in 1955. *Source* personal collection

But, as my parents became more at ease and settled into their lives in Britain, less absorbed in managing the day to day, they felt able to reflect on their past, more frequently revealing snippets about their lives prior to arriving in London. And I began to listen, with care and respect, subsequently encouraging them to recollect episodes from their enthralling adventures, asking questions, prompting and reassuring. In Gujarati there is a term for a listener of stories, *sambla*, one who imbibes the tales recounted to them, who acknowledges that the memories of the teller are necessarily entangled with their own and who will subsequently be shaped by them. Gradually, as their stories emerged, my negative thoughts were replaced by a wholly different image of my parents. They were far more lively, modern and exciting than I had imagined. My sisters and I would repeatedly ask them to tell us the story of why they decided to undertake such an adventurous road trip only so that we could relish in hearing our parents' love story. Mum would readily submit to these requests, delighting in the opportunity to remember my parents' youthfulness: 'we were just ordinary people, we hadn't seen places, but we were adventurous. We wanted to see the world together. The bits I remember of that long journey are love fragments. We stood side by side on a bridge or a hill looking around us. Just sharing the view. We were in love'.

Unlike Irene Kacandes (2009), whose father was unwilling to speak about his traumatic past and who therefore had to rely on gathering snatches heard from behind closed doors, invitations to my parents to speak about their travels were met with enthusiasm. Theirs was largely a joyful account. Eagerly and generously, they began to share their stories, relishing the opportunity to reminisce together. Their tales seeped out gradually, slowly unfolding over many decades, nourishing our family history and transforming who we thought they were. They would recount anecdotes of their trip at family gatherings, or when an incident, person or object sparked a memory. The telling of some stories was prompted by current events and news reports. For instance, the imposition of stringent immigration controls and the refugee crises in Europe caused them to reflect on, and appreciate, the ease of movement they had enjoyed in the 1950s on account of the Commonwealth passports with which they were furnished.

As the stories of their travels unfolded, it became clear that their desire to travel to Britain, their expectations and aspirations, had been profoundly shaped by having absorbed the stories of others.

Growing up in India under colonial rule they had imbibed powerful idealisations of Britain as a more cultured, orderly, even glorious world. Like other colonised people, they learned about Britain through the stories that swirled through their education, and in literature and films, that conjured up a nation steeped in myth and tradition and that saturated the romantic images of landscapes and historic places that adorned the picture calendars that hung in their homes and schools. These colonial representations purveyed a selective knowledge of Britain, instilling an imagined geography in which the metropolitan, imperial space was the epitome of civilisation and progress.

As a consequence, these fanciful colonial imaginaries, so powerfully fabricated through the stories in which they had been enmeshed as colonial subjects, nourished their Anglophile tastes and aspirations and promoted a profound desire to visit Britain. Indeed, actor Art Malik (2016, p. 1) remarks that he would always respond to the question of why he came to Britain with "Well, because when you [British] came to India, you told us all that it was the most brilliant place on the planet". So all-consuming were these fantasies that even anti-colonial nationalists, like my parents, were motivated to visit the centre of empire. Indeed, as Mohammed Rahman (2009, p. 130) writes, "growing up under the wings of Englishness in the colony, there were hardly any Indians, especially English-educated Indians, who had not longed to travel to Britain".

With increasing opportunities for travel following the end of the Second World War in 1945 and the advent of Indian Independence in 1947, those with the desire and means could now fulfil their long-standing aspiration to experience at first hand places with which they had become vicariously familiar through the stories they had been told. These accounts of an imagined, magnificent Britain influenced my parents to undertake a trip, embarking on their journey projecting an overwhelming and unquestioning admiration for Western civilisation. Their absorption of the flows of ideas, values and representations from Britain to India had cultivated a deep awareness of celebrated histories, heroes and iconic sites. Thus, travelling to Britain did not seem like a journey into the unknown but more as confirmation of what was already foregrounded through the dissemination of these dominant, authoritative stories. In some ways, their travels to the imperial heartland constituted 'a pilgrimage' with ideas about Britain having taken "root in the Indian collective imagination, even before the travellers had set off on their journey" (Rahman, 2009, p. 133). Consequently, when they arrived, their first impressions of Britain were "tainted with a sense of déjà vu" (Mukhopadhyay, 2002, p. 293).

As they drove through Britain, their tourist gaze was, at least initially, profoundly shaped by these colonial stories. They visited renowned historic locations and drove through scenic landscapes. They were keen to spend time in Oxford and Cambridge as centres of British educational excellence and were drawn to stop at seaside resorts, cathedrals and castles that had infused their imaginations at school. These visions were further fostered by their rather hyperbolic guidebook, *Britain: A Book Which Attempts to do More Than its Size Permits* (1952), that whetted their appetite in claiming that Britain was the home of the touristic sublime: "a country which by reason of its structure, position and history has, perhaps, more beauty, interest and variety to offer the visitor than any other country in the world" (BTHA, 1952, p. 7).

Besides accounts of the glories of Britain, there were other forms of storytelling that influenced my parents' itinerary. Some of the most significant destinations on their trip were symbolic sites associated with English literature. Desires to visit these places were stimulated by their deep familiarity with the work of iconic British poets, playwrights and novelists whose writings had been compulsory reading on their school curricula. The Longman literature reader that they had pored over instigated their urgent priorities to go to Stratford upon Avon, the birthplace of William Shakespeare, visit Lord Byron's estate in Newstead and drive to Edinburgh because of its association with Walter Scott. Besides their acquaintance with these classic writers during their colonial education, the flow of English language books across the empire meant they also had access to a host of popular British novels and had developed an appetite for the works of George Orwell, Graham Green, Somerset Maugham and the globally popular Agatha Christie. Consequently, as B. Venkat Mani writes in *Recoding World Literature* (Mani, 2016), through this 'bibliomigrancy', literary works migrated to different parts of the world, shaping local reading practices and instilling ideas about what constitutes a 'good story'.

The deep influence of these British literary works seemed to fulfil the imperatives and beliefs articulated by the architect of Colonial Britain's Educational Policy in India. In a speech at the Edinburgh Philosophical Society in 1846, Thomas Macaulay, a historian, poet and politician, offered a toast "To the literature of Britain ... which has exercised an influence wider than that of our commerce and mightier than that of our arms" (quoted in Whitehead, 2005, p. 319). He famously declared that "a single shelf of a good European library was worth the whole native literature of India and Arabic". My parents' stories of their pilgrimages to locations associated with literary works, places in which writers lived and worked and wrote about, reveal the success of these colonial policies.

Imbibing these stories from elsewhere, that were printed in books that moved from one geographical area to another, shaped my parents' knowledge of British culture and language and played an inspirational and transformative role in their travels. Their tastes and judgement, influenced by well-worn colonial stories about what constituted 'good' literature, were entwined with a valorisation of the English language. Indeed, my father's archaic vocabulary, style of speaking and turns of phrase, were heavily influenced by English literary works that incited a lifelong aspiration to 'master the language'. He ignored our hilarity at his frequent malapropisms and cared little when we corrected his mispronunciations, ceaselessly employing any new words he came across even if they were not precisely applied.

For some time after they set off on their road trip in 1955, they held on to the colonial idealisations that had shaped their expectations of Britain. However, as they encountered the 'real' Britain, they became less likely to simply ingest what they observed and sensed without reflection and judgement and began to author their own stories. Inhabiting British space, they became increasingly discerning, disentangling and disaggregating their colonial influences and imagining possibilities other than those that had become part of their more sequestered worldview. At first, these shifts in their thinking about the centre of empire were imperceptible as they resisted the instability and discomfort that accompanied the realisation that the stories they had been told and the realms they had envisaged were partial and illusory. Gradually gaining confidence, as a distorted colonised lens was supplanted by an unfiltered gaze, they began to read Britain otherwise, challenging how it had been presented to them before they had arrived. Encountering Britain no longer affirmed their colonial imagination but enfolded them in a myriad of new sensory and affective affordances that circumvented dominant colonialist and nationalist narratives. It became clear that the imaginaries they had imbibed through the stories they were told were not immutable and singular but were always in danger of being revealed for the over-inflated myths they were; Britain was other to how it had been storied.

My parents began to create their own 'counter-stories' fuelled by a measure of critical reflexivity as their search for the British sublime eluded them; their idealised visions of Britain as a more cultured and orderly world were unsettled, rebuked, disrupted and reinterpreted. They told tales about a multiplicity of everyday, ordinary and extraordinary features, behaviours and places which deviated from the familiar representations that had been reiterated at their schools and in the popular English language films, books and magazines they had enjoyed. Profoundly disorienting for them was witnessing poverty and a post-war Britain of dereliction, ruins and bombsites. More troubling, as they wandered through Britain was their awareness of being noticed as different. Though frequently welcomed, and with hospitality extended to them, at times seen as exotic and greeted with a naïve curiosity, they encountered others who invoked racist attitudes, drawing on well-worn hierarchical distinctions in which my parents were regarded as inferior.

However, although they developed their own distinctive sensibilities about the country, and in so doing confounded colonialist accounts, their new postcolonial stories revealed how racist, hostile incidents were often shrugged off: 'some people were not so nice but why should we remember those bad things, it hurts the heart' they would say. When pushed to recall an unpleasant encounter, they would reprimand me with 'the heart is too small to hold negative thoughts and bad feelings'. My father emphasised 'People generally were nice, very rarely we met people who were not nice and if you want me to remember them, that will take me a long time. Even those who for some reason or another were against us and our way of life, you didn't hate them'. Yet, although their stories were dominated by delight in what they had encountered, some antagonistic confrontations were recounted, often tearfully, with pain and sadness. I have heard, more than once, their encounter with racist nuns and of being taunted by a group of young men. While some upsetting events tended to be communicated with humour, not allowing them to 'hurt the heart', the fact that these and other malign episodes were recounted so regularly testify to their deep, long-lasting, heartfelt impact. A reminder that it is not only the content of a story that is relevant to understand its power, but the spaces, absences and repetitions in their telling, alongside the cadence, emotion and tone of the storyteller.

In his semi-autobiographical novel, *Black Shack Alley*, Joseph Zobel (1980) asks what makes a story worth telling. My parents' tales were belatedly told and only within the confines of our domestic space. They could have remained wholly private, evanescent tales only shared among family members. Yet as their stories multiplied, it became clear to me that this was a collection of stories that was not only worth telling but demanded to be told. When contemplating what is remembered or forgotten, what belongs or is left out, what is considered important enough to record and preserve for posterity, it dawned on me that the unique episodes that accumulated contribute a distinctive voice; stories that need to be heard and recognised. As artist and cultural theorist Ross Gibson (2015) writes, while stories are often discrete and intimate, they can be used to provide an alternative, critical lens to explore histories unfolding in particular settings and across the larger world. I am now in the process of bringing together my parents' stories, to narrate their experiences. Together, their stories of an exceptional, remarkable road trip, in which they detail their desires, itineraries, adventures and perceptions veer away from colonialist accounts and the ubiquitous travel stories of European and American writers, by offering an alternative, rare account of non-western travellers. In addressing what Saidiya Hartman (2008) refers to as 'archival silence', wherein certain stories are destroyed or never uttered in the first place, my parents' stories identify what is missing from our collective histories.

Edward Said (1978) wrote that the formerly colonised must confront their entanglement in processes through which they have become defined and confined. Prevailing, authoritative accounts of colonised people were often told not by them but incorporated into colonial narrative forms, disseminated as folk tales, dispassionate scientific reports and fables that identified the archetypal characters of another culture. The unrepresented, unarchived stories of my parents could not trouble the wider world, although their preoccupations, telling of incidents and opinions would, importantly, have veered away from these worn-out, sterile colonial stereotypes and tropes. That I can now more publicly retell their stories signifies that things are changing. Yet, I am conscious that access to storytelling has always been uneven across history and culture, and that the opportunity afforded to me, and the ability to write this account, has very much depended upon what Lynne Pearce and Nicola Spurling (2024) refer to as "a complex nexus of historically-specific factors: social and economic privilege, geographical location, and available discourse". I am fortunate to be able to narrate their stories, yet my privilege does not detract from the critical contribution I am able to make by inscribing a previously marginalised story onto the public record.

Recounting my parents' story, however, is not straightforward; it is inevitably complex and complicated. How to tell their story in a way that conveys the intimate and evocative qualities of my parents and their travels yet goes beyond a personal account to one that resonates more widely relies on the act and art of narration. As I attempt to weave together what began as sporadically told personal stories over an extended period, I realise that creating a narrative from such fragments requires consideration of the workings of memories, how stories are told and how they change over time. My parents selected the events that they narrated; they implicitly interpreted the past by what they omitted and what they emphasised, which incidents and experiences they embraced, and which remained hidden. Their memories are emergent; they are not passive depositories of facts, but lively practices through which feelings, sensations and meanings are imparted. Although certain episodes have become polished with repetition, most stories were rarely related consistently and changed shape, emphasis and focus over time. With advancing age and temporal distance, rosy anecdotes and romantic recollections began to dominate the retelling of their travel adventures.

And, parents grow old. By the time I fully appreciated the significance of my parents' story and embarked on researching the details, their memories were fading. One day when my father was unable to respond to my questions he said: 'I am sorry *beta*, it is the memory loss. I have lost sight of things and matters; I am not able to tell you details. You are collecting these stories too late'. This dwindling capacity to recall is not only a consequence of ageing but a broader characteristic of how we recollect. As Carmen Maria Machado (2019, p. 271) writes, it is common that "once you are departed from an event, some essential quality is lost forever". In my parents' case, this loss could at times be repaired since they often remembered different details, filling in the gaps in each other's memories and in so doing creating a combined recollection.

The relationship between tellers and listeners also shapes the stories that are told, how and when, and what is remembered. As a listener, I draw on my long and intimate relationship with my parents and our deep emotional connections. And so, where there are absences, I have conjectured about what my parents might have thought or felt. Like Maria Stepanova (2017, p. 29), I too disclose that in "place of a memory I did not have, of an event I did not witness", I am sharing interpretations of my parents' recollections.

In illuminating voices and accounts that have largely been excluded, hidden and overlooked, and inscribing these previously silenced stories on the archive, my parents' account responds to Johny Pitts (2019) encouragement to flip the record over and listen to 'the B side', to the stories outside the stories. My parents' stories are about storytelling and memory, of colonialism and postcolonialism, of global encounters and affinities and ultimately, of small people making their own modest contribution to big history.

In gathering up those fragments that have resonated through the years, I hope that my parents' stories, in which meanings and feelings implicit in the past travel into the present, stimulate new and different kinds of imaginings.

References

Behrendt, L. (2021). *After story*. University of Queensland Press.

British Travel & Holidays Association. (1952). *Britain: a book which attempts to do more than its size permits*. The British Travel and Holidays Association.

Gibson, R. (2015). *Memoryscopes: remnants forensics aesthetics*. Apollo Books.

Hartman, S. (2008). Venus in two acts. Small Axe: A Caribbean Journal of Criticism, 12(2), 1–14.

Kacandes, I. (2009). *Daddy's war: Greek American stories*. University of Nebraska Press.

Kureishi, H. (2014). *My ear at his heart*. Faber & Faber.
Machado, C. M. (2019). *In the dream house: a memoir*. Serpent's Tail.
Malik, A. (2016, March 13). *Television has gone backwards - it's no longer colourblind.* Radio Times. Accessed March, 2025 from https://www.radiotimes.com/tv/drama/art-malik-television-has-gone-backwards-its-no-longer-colourblind
Mani, B. V. (2016). *Recoding world literature: libraries, print culture, and Germany's pact with books*. Fordham University Press.
Mukhopadhyay, T. (2002). *A visit to Europe*. Gale and the British Library.
Pearce, L., & Spurling, N. J. (2024). Auto/biography and mobilities in the time of climate emergency. *Mobilities, 19*(5), 807–822.
Pitts, J. (2019). *Afropean: notes from Black Europe*. Penguin UK.
Rahman, M. A. (2009). *The Gaze: imperial Britain in Indian travel narratives. 1765–1947*. (Doctoral Dissertation,. The University of Manchester.
Roy, A. (2006). *Indian artists inspired by English Countryside.* Accessed March, 2025 from https://www.telegraph.co.uk/news/1528549/Indian-artists-inspired-by-English-countryside.html
Said, E. (1978). *Orientalism.* Pantheon Books.
Stepanova, M. (2017). *In memory of memory*. Fitzcarraldo Editions.
Whitehead, C. (2005). The historiography of British imperial education policy, Part I: India. *History of Education, 34*(3), 315–329.
Zobel, J. (1980). *Black shack alley. 1950.* Trans. Keith Q. Warner. Lynne Rienner Publishers.

4

Haunting: Storying Lively Death

Sarah Wright

Abstract This chapter follows a story of a cemetery as a place where life and death, play and loss, order and transgression, the material and the transcendent are deeply entangled. Cemeteries are more-than-human and multiply storied places, places that themselves tell stories of lives cut short, and of those who remain. They are at once adamantly material, a place of bodies, tombstones, earth and decay, and extra-ordinary, haunted by death, loss and otherworldliness. This chapter looks to the ways that cemeteries haunt, refusing the monologue of one-way storying. Cemeteries defy ideas of purity and stasis, and instead bring into focus a multi-layered, multi-directional co-storying that has the potential to make and unmake worlds differently, though always in limited, or incomplete ways. They bring into the present that which might otherwise be seen as gone, including many uncomfortable truths and unresolved violences of colonisation. In this chapter, I invite a haunting, a re-membering, through which many of the cemetery's more-than-human beings, including its graves, memories and spirits, demand attention to their ongoing presence. They call me now and then. They insist that I listen more deeply, and live in ways that acknowledge the cycles past,

U. Kothari et al., *Stories of Place* https://doi.org/10.1007/978-981-95-4921-4_4

those that have gone before, and the connection of earth, sea, sky, life and death.

Keywords haunting · storytelling · death · place · more-than-human agency

When I was a child, we used to play in the cemetery. It was right near our house, just on the other side of the electricity substation. This was and is Garigal Country, now an outer suburb of Sydney. The house I lived in growing up had once been a farmhouse on several acres, well removed from the city. At some stage, most of the block had been sold to create the substation and, as no one was allowed to build right next to the substation itself, there was a band of grass and open space that acted as an exclusion zone around it. That was the area we played in, walked the dogs, rode horses, played handball or rode billycarts; there were market gardens on all sides and the cemetery which, at the time, was marked off by a low, dilapidated wooden fence.

This landscape is the landscape of my childhood. It's a story-place that lives somehow deep in my being; something that is me, that makes me, a re-membering of sorts (Grande & McCarty, 2018). As a place, the cemetery is both magical and, in some ways, grim. The substation giving off radiation, the cemetery itself with its grief and physical reminders of death, part deathscape, part wasteland, part wonderland. A haunting emerges on multiple registers, through memory and re-membering, through present-absence, in the spectralities and materialities of deathscape (Maddrell & Sidaway, 2012).

It's hard to trace, too, who is really doing the telling here. It's me, my memories, my relationships, the cemetery itself and the substation, ghosts and the unseen, different pasts and histories that jostle the present. These places, the cemetery and the substation, in a very tangible way, make up the very me that tells. They grew me up, along with my (human and non-human) family and friends. And, more, they tell stories too, stories that refuse a monologue or simple understanding. I wonder if the substation had any role in my brother's childhood cancer. I recall with rose-tinted nostalgia playing Star Wars with him in the oversize gutters that rimmed the factory: how the reddish-brown surfaces would stain our hands. The place, with its rich be(ing)longing that sparks in/as me, is thick, multi-sensorial, complex, shared.

And so, in this chapter, I tell a story of re-membering place, of place re-membering me. In contrast to dis-memberment, re-membering is a process of being put back together, remade through an affective experience of retelling and memory (Daley & Wright, 2022; Grande & McCarty, 2018; Smith et al., 2021). I yarn up a cemetery and some of the many ways that it haunts, and, in this haunting, tells stories differently. I mean to speak of death but, as Virginia Woolf points out, life keeps breaking in as usual (Woolf, 1922, p. 42, cited in Lorimer, 2019). This chapter is a place-writing, a love-letter written by me to place, by place to me (Lorimer, 2019). And it is an effort for me to learn to be haunted, to listen to a haunted place, to story and be storied by it.

As a child we'd walk past the substation and jump the fence to the cemetery; sometimes it was with my brother or sister, or with my best friend, Alison. I think it was mostly Alison and me who were into it, actually. We used to spend hours there. We would tend grave sites, idly wandering down the aisles reading headstones, or we'd leap over the graves. I remember one time, Alison and I wanted to see what it looked like inside the men's toilet block, so we got dressed up, putting on a beanie hat to hide our hair, and went in to check it out. We put coins on a grave to try and see if we could be visited by a ghost.

(See (Fig. 4.1))

The cemetery, for us, was a lively place of play and transgression. Yet, there was order here, and grief; the presence of death of which we were in no way oblivious. We were earnest as we tended graves and read inscriptions, pausing in respect of someone lost, especially if they were young. And we were playful, wayward, jumping graves, running amok. It's fitting that this was the place we chose to investigate the mysteries of a male public toilet. We actively encouraged a border-crossing between worlds, inviting spirits to rise, seeking companionship with the dead. In these ways, the material and the transcendent, presence and absence, play and solemnity were always deeply, yet prosaically, entangled.

Fig. 4.1 Re-membering my ghostly self-playing in the cemetery. Credit: Sarah Wright. Image of the toilet block at Mona Vale cemetery courtesy of Northern Beaches Library local studies

This sense of deathscapes as places of in-betweenness and liminality is found in many discussions in cemetery studies (Rugg, 2022), necrogeography (Semple & Brooks, 2020) and hauntology (Holloway & Kneale, 2008; Stevens & Tolbert, 2018). The presence of death and haunting appears to encourage an orientation that "gesture[s] towards that which is neither present nor absent, nor dead nor alive (Davis, 2005; Luckhurst, 2002; Roberts, 2013)" (Stevens & Tolbert, 2018, p.28). The ghosts are vestiges, spectral presences that linger, bringing into the present that which might otherwise be gone, making their presence felt in plural ways (Edensor, 2005, 2008). As Edensor (2008, p.327) elaborates, ghosts reside in our present selves, in our memories, as residues of distant places, relationships and times. And so it was for us at the time, and now, in the retelling, this re-membering of self and place.

I cared about that place; I felt like I had a direct relationship to it. It was a place of calm and of freedom where the rules didn't quite hold the same way. It was a place of both safety and independence, of adventure, and yet a place where I had a clear role to tend others, self-identified though it might have been. I think mostly it gave a sense of another world, another dimension, something very *other* that we would find out about one day. Yet it was a place of the day-to-day too, not separated from play or exploration or from life, such a matter-of-fact relationship we had. Perhaps we sensed that, in our childlike pragmatism, we knew something about the place and the people and life and death that adults had lost touch with. We were drawn into something that felt intriguing, a little magical maybe.

I don't think my parents knew we were there. It's not like they forbid it, but I think we felt instinctively that it might be better just not to mention it to them. So, although in many ways playing in the cemetery didn't seem at all strange to us, we knew it was transgressive of course. It wasn't quite like playing at the old outhouse where we knew there were venomous spiders or like walking on the top of the neighbour's fence; both those places were explicitly forbidden. The situation with the cemetery was more ambiguous. As Annabel Deering (2012) elucidates in her thesis, *Over their Dead Bodies: A Study of Leisure and Spatiality in Cemeteries*, it was an ambiguous place, something 'other', heterotopic in expression "through the simultaneity of different forms of time, through the juxtaposition of life and death, art and nature" (p.10). Cemeteries with their haunting defy a clear separation between the past and present. Indeed, the present becomes an expression of a living past in which the realities, stories and relationships of now co-become with the realities and relationships of those that are gone.

And, in my memories, the cemetery *feels* different. It was somehow an intimate relationship we had, with the graves, the flowers, the grass, the trees, the bark, the cold, the warmth and the heat. And intimate with strangers too that had passed away. I didn't know anyone who was buried there, and there weren't often burials as the site was pretty much full. In my memory we very rarely encountered another person, another person who was alive, that is. We roamed and we cared in our childlike way, though I suppose from the outside we probably looked like feral kids. I imagine we'd have been chased out as troublemakers if someone saw us.

Stories were told differently there, in ways that actively invited a boundary-crossing, between human and corpse and decay, between human and spirit, between permitted and not, between sacred and mundane. And the line between story-tellers and story-listeners blurred too. We listened for ghosts, we were told by grass and tree, our coins on the grave were an invitation to connect. Our listening was sometimes felt in the prickling sensation of being in the presence of something other-worldly, stories heard through hair raising on our arms, on the compulsion that would come on us occasionally, to run home as fast as possible.

There was care there, for death, for unknown others, people that we never knew, could never know, except we *did* know them in some way, in the care we took as we tended their memory, their ghosts, their earthly remains. We particularly cared for children's graves. As children ourselves, we connected with those children and their remains deeply. We felt a solidarity, a grief that their young lives could be cut short, perhaps made all the more poignant by the teaching they offered us that we, too, could die as children. We would seek out children's graves and clean them; if there were flowers or toys, we'd prop them up if they had been knocked over. We would even redistribute flowers if we felt it necessary; that is, if we saw one grave with what we considered an overabundance of flowers while a child's grave had none. I don't remember being, in general, overly afraid or creeped out; it didn't seem ghoulish or dark, or weird. Rather, even those times we invited in horror, telling scary stories or daring ourselves to go at night, it still seemed matter of fact, a part of life, a good place to be.

We were building relationships with the past, past people and places, and with mortality, with the unknown. Francis et al.'s (2005) exploration of the socialities of London cemeteries looks to graves as domestic spaces where everyday behaviours such as gardening are a form of ongoing dialogue with and care for the deceased. In our lively deathscape, our solicitude and dialogues were acts of relationship building, new relationships rather than ones that began in the lives of those now passed on. And there was ritual too, as we solemnly laid flowers and otherwise expressed our own connection with the afterlife in ways that diverged from formal, modern expressions of grief and mourning (Davies & Rumble, 2012; Jedan et al., 2020).

The materiality of the graves, the flowers, the bodies, the stones with their wildly abbreviated stories of life, all facilitated communication between us and the dead (Francis et al., 2005). Those gravestones told thousands of stories in few words, and in words not said, in what we could read in the dates of birth and death, the size of the mound, the flowers and mementoes left (Lorimer, 2019). Here, a loving brother, father, grandfather and son, gone but not forgotten. There, a family reunited. Some are with God. Some with each other. And here, a small unloved-looking mound, words rubbed off by weather and time. We felt the past, and we'd wonder about the stories behind the dates and the short words on a headstone. We tried to read behind, beneath and through.

We did acknowledge and sense the sorrow there; we grieved for the children with their small graves. There was a connection to things that had gone. This place was life and death, suggesting a kind of permanent impermanence. No one was exempt, not the rich with their fancy tombs, nor the seemingly forgotten with their untended simple plates in the grass. It was a place that linked us both to the past and also to the future; a place where we are all headed, one way or another.

It was not just people, not only human bodies, that storied here, whose materialities made place and made us and which I remember now. The stones themselves, the cracked concrete of a tomb, the faded plastic flowers. A heart-breaking teddy bear. And the feel of the place itself. I remember the trees, down the side, in particular. I guess these beautiful old trees were remnant bushland. I mean, there was bush up behind the market gardens, and some old trees, three in particular that were often witches in our games, left for some reason in the grass of the substation, but the trees of the cemetery were a beautiful, old, wide strip. We'd sit under them, get bitten by ants, the crunch of dry leaves digging into our legs, and their presence held us. I think now about how much they'd seen.

Some of those trees likely predated the formal cemetery. I looked up when the graves first went in, and the cemetery was gazetted in 1905. Its first grave was for a two-year-old named Percy Jonston. I can't remember if we tended that grave but I think it was unlikely. We mostly were concerned with those that seemed forgotten, by those that seemed less "grievable" in Butler's (2004) terms. Ungrievable lives, for Butler, are those that are not, cannot, be mourned as the lives were not seen, recognised, valued, acknowledged as worthy in the first place (ibid). I wonder now if there were older burial sites there that predated the gazettal as well, ones erased by the cemetery itself.

Those that were forgotten or dismissed haunted us, and they remain an important presence as I remember now, meaningful in different, perhaps deeper ways. This is a different layer of the haunting that we glimpsed as children as we searched for poorly tended, and marginalised graves, and noted the fancy tombs on the one hand, the humble plots on the other. In this and other ways, cemeteries reflect and prolong, in death, differences and marginalisations of life (Maddrell et al., 2021). Such are the realities of necrospatial marginalisation (Rugg, 2022) that go deeper, are vaster than we recognised at the time, even as absences scratched at the edges of our consciousness. The ghosts excluded or demoted by virtue of faith, spectres signalling the ways some deaths, some peoples, including those who were insane, indigent, imprisoned, who were enslaved at the time of their death, or the victims of genocide needed to be heard. There were and are stories cut off in premature death, silenced and hidden, ignored and wilfully, violently silenced. So many exclusions, so many untended ghosts, a deep layer of haunting, of dismemberment, rather than a re-membering.

Many scholars have spoken of social haunting, drawing particularly on the work of Avery Gordon, that speaks to the shadows of social life, of lives and deaths ignored, marginalised or violently erased particularly those taken as a result of slavery, colonisation and their lingering afterlives (Best & Ramírez, 2021; Gordon, 2008, 2011). Haunting, according to Gordon, points to "how that which appears to be not there is often a seething presence" (2008, p.8). Ghosts interpolate, refusing to be ignored, bringing attention to unresolved violence, particularly when these structures of violence and oppression are purportedly things of the past, without relevance to the present. This, as Best and Ramirez point out, is the way of ghosts (2021).

Such hauntings pervade planning, property regimes and all the structures of settler-colonial life. While sometimes hidden, they nonetheless seethe. Amaka Okechukwu considers *Rest in Peace* murals in Brooklyn, New York, forms of public art that commemorate lives of the deceased pointing to the way these make visible unresolved violence of Black dispossession and disposability. For Okechukwu, this is "wake work" (Sharpe, 2016) that speaks back to the enduring violence of lives ignored, lost and stolen, a social haunting that announces that what appears to be absent "is actually here, at this time" (Okechukwu, 2022, p.139). The murals act as archives for collective memory, calling for community members to remember the dead, honour their lives and their deaths, and urge those still alive to keep living despite ongoing trauma (Gordon, 2011; Okechukwu, 2022).

Many ghosts reach over distance to tug at me now. I am aware of social hauntings that call for a different kind of re-membering place, including, centrally, the ways the cemetery plot, and the house I grew up in and the substation and, indeed, the whole of Australia, is Aboriginal land, is made up of land that was and is stolen, full of stories with a much longer history of care and belonging. These are different kinds of hauntings associated with the cemetery: as we walked to the fence and climbed it, as we played hopscotch on the road leading to the building itself, we were walking, playing, sitting, on land that had prior belongings that had been ignored, denied and dismissed through the systems of possessive, capitalist and white-supremacist extraction associated with settler colonialism (Moreton-Robinson, 2015). Belongings that we didn't glimpse then, but that reach out to me now. They reach out to me through the powerful sharing of Indigenous academics (Kwaymullina, 2020; Moreton-Robinson, 2015; Watego, 2021), and of loved ones and mentors including my Momus of the Bawaka Collective (Bawaka Country including Suchet-Pearson, 2020) and Yandaarra (Yandaarra with Aunty Shaa Smith et al., 2024).

As Goenpul scholar Aileen Moreton Robinson (2015) points out, throughout Australia, systems of Western property, including the many different tenures that make up urban spaces including private property, crown land and public spaces such as cemeteries, seek to annihilate and replace systems of Indigenous Law/Lore and belonging. Part of this process is to story the lands, seas, skies and the deep and complex spiritual, social and cultural relationships of Aboriginal Country as empty, as Terra nullius, belonging to no one in ways that forcibly ignore Indigenous peoples, laws and cultures, even the agencies of Country itself. This is how land may be claimed by the Crown, gazetted as a cemetery without reference to those that live and died there. It is how a new death-city can rise from a 'vacant lot' making settler-occupation appear unquestionable, inevitable (2015). In this way, the presence of settlers in life, and in death, including this magical and mysterious cemetery of my childhood, is established and maintained through an ongoing erasure, a necropolitics that seeks the death of certain people, culture and Country (Haritaworn et al., 2014; Mbembe, 2020). Such erasures and violence haunt. And, through this social haunting, the cemetery asks, whose death is memorialised there? What prior belongings haunt the orderly graves? What spectres whisper through our childhood play, what other re-memberings need to be tended?

So, at the same time that absence is conjured, connections nonetheless linger, in the ongoing presences of Aboriginal people and culture, in Country and its spirits. Such presences cannot be erased or relegated to the past. As Moreton-Robinson goes on to point out, Country is never nullius, neither in its liveliness nor in what comes after, and possession is never complete, never done, never settled (Moreton-Robinson, 2015). This too is a haunting, a reality that unsettles the settler state, that brings uncomfortable truths and insists on different responses. Aboriginal Country, a concept that goes far beyond land to include all beings and processes, the skies, spirits, waters and dreams, always was, and always will be, alive with the spirit of those who have passed away, human and non-human.

And so, as I remember place, the playful places of the cemetery and the lively death it brings and brought, it becomes clear to me that the spectres and hauntings of the cemetery are myriad. There are many ghosts there, living still in the soil, the plants, the air, in memories and in dreams. These are hauntings that speak of things denied, ignored and suppressed. And there are nourishing hauntings too, in the playfulness and transgression that I was taught, and in the presence of different kinds of spirits, different pasts and the agencies of Country. There are Ancestors, Old

Fellas, beings that live in/as Country, including in/as the precise Country that is the cemetery that I story and that stories me.

Indeed, for many First Nations people, Ancestors are an important presence, a different kind of haunting, a haunting in the most nourishing sense (Bishop, 2023; Bishop & Tynan, 2022; Cavanagh & Standley, 2020; Shaw et al., 2006). Where I live now, many years later, on Gumbayngirr Country about 6 h drive north of the places of my childhood, I have learnt deeply from Gumbaynggirr Elders, mentors, friends and loved ones including Aunty Shaa Smith, Uncle Bud Marshall and the big mob that is Yandaarra, a Gumbayngirr (Indigenous people and Country of the mid-north coast of NSW in Australia) and Yiralli (whitefella, non-Indigenous person) research collective of which I am a member (Wright, 2024). Aunty Shaa and Uncle Bud point out that Ancestors, the 'Old Fellas', are here, watching and guiding, and they need to be treated with respect (Marshall et al., 2022; Smith et al., 2020; Yandaarra et al., 2022). Ancestors have stories to tell, wisdom and law/lore to impart. They can guide respectful action and can pull you up if you are doing the wrong thing.

Ancestors live in and as place. They are the land, they are Country, and they are, in very material ways, part of us. Aunty Shaa Smith and Yandaarra suggest a need to understand re-membering as embodied and more-than-human. Aunty Shaa talks about literally putting ourselves back together with and as Country, "to wake up this memory in [our]selves, of how to live in balance from a place of connection, of wholeness" (Smith et al., 2021, p.7; see also Grande & McCarty, 2018; wa Thiong'o, 2009). Asked about how to come into good relationship with Country, how to listen and learn, Aunty Shaa, storyholder for Gumbaynggirr Country, speaks with her daughter Neeyan about the importance of Ancestors and their teachings (Yandaarra with Aunty Shaa Smith et al., 2022, p. 719).

Aunty Shaa: For me it's been communicating with my Ancestors and opening my mind up to dimensions where we meet and communicate, that ground; learning about what the protocols are for that … and it's all done by intuition and sensing; being tuned in to vibration and frequency; and then it's understanding that the structure in that place cannot be destroyed.

Neeyan [Aunty Shaa's daughter and co-founder of Yandaarra]: It's about believing our Ancestors are still with us. They're not gone; that feeling of them being here becomes real too and that relationship becomes stronger.

Aunty Shaa: The more time I spend with my Ancestors, the more in-tune I've become and they just help so much … you're walking with them … Our Ancestors are like working their butts off to create this consciousness around oneness and relationships.

Here, haunting becomes a positive and nourishing form of guidance that occurs in deeply embodied ways; something, as Aunty Shaa says, that cannot be destroyed. Direction from Old Fellas, who are deceased but who retain a formidable presence in/as the land and in/as spirit, 'working their butts off' is all-important. For Yandaarra, we hold that "listening and learning from the Old Fellas, listening and learning from Country and following Gumbaynggirr protocols are at the heart of how we approach research. Old Fellas and Country guide us, they teach us, they help show us what to do" (2022, p. 710).

So, death teaches and Ancestors haunt in very real ways, well beyond a metaphorical understanding. Being haunted in such a way resonates with Birrbay academic Arlene Mcinherny's (Mcinherny, forthcoming) appeal to listen "with the whole body, including the ancestral body, which remembers rhythms long buried under colonial noise". Haunting guides right action, encourages good relationships, allows for positive co-becomings held through the co-becomings of non-linear time (Bawaka Country et al., 2016; Daley & Wright, 2022). It follows that, while as Stevens and Tolbert (2018) point out, haunting as understood in the social sciences, is often treated as a metaphorical tool, even a fashionable gambit, for Yandaarra and many others, the presence of Ancestors is real; Old Fellas communicate through intuition, dreams, and through Country; they are materially present in/as the land and its beings as well as in a person's memory and genes; they are tangible, relatable. They haunt. To attend to Ancestors is part of what it means to really listen to stories in embodied ways, and to story life well.

So, that leaves me in a place of actively inviting a haunting. I am there, once again, in the cemetery, reaching out to the dead, the unseen, the unknown. I make tentative moves to come to the edge of who I am (now), what I know and what I don't, and, from my place, to care across boundaries; to dwell in liminality. What would it mean, to haunt and be haunted, in some of the many ways, on the many levels, that one might? Asha Best and Margaret Ramirez ask, "How might we think of haunting as a mode of refusing displacement, banishment, and archival erasure as a way of imagining 'livable' urban futures in which Black life is neither static nor obsolete?" (2021, p. 1043). Here, in the past/present/future of the cemetery, in Australia, this means a much-needed unsettling of "the taken-for-granted privilege of settler dominance in Indigenous domain" (Howitt, 2020, p. 193). Re-membering might orient me past the intended severing of coloniality towards a revisioning, bringing the elsewhere, the other-where close (Grande & McCarty, 2018; wa Thiong'o, N., 2009).

And I commit to responding to this call, to playing close and careful attention to spectral invitations and interpolations. Haunting might help me think and imagine differently, yes. And to live and be and remember differently also. I am haunted by the past; the ways my present is empowered by the violence done to people and to Country. I am haunted by futures foreclosed by the violence of the present. Haunted by Ancestors, my own and others, who might nourish, who might help guide me differently, towards a respectful relationship premised not on possessive belonging but on a tentative, differentiated coming-together. Haunted by present absences of people, non-human beings, those that should/could have been there, that which seethes beneath the surface. Haunted by embodied messages, of Country, by rich soil created through the decomposition of others, by plants and fungus that grow on decayed life, cycling through generations, transforming, being transformed, in my body and from there, through death, onwards, my inevitable material kinship with soil, ashes and dust that will be my mortal remains. Haunted by death of myself, my loved ones, those I never knew, by the realities of my permanent impermanence. Haunted by joyful memories of myself and place and our co-becomings across and as time and place made possible only by the absences and dispossessions of others. Haunted by the landscape of now and then, and the poignant possibilities of re-membering, putting myself back together with and as place.

I ask ghosts to subvert narratives, to allow their voices to be heard, to speak loudly of Indigenous and subaltern realities, of those marginalised (Cameron, 2008; Lloyd, 2005; Roberts, 2013). These hauntings are stories, they make stories, they story place as place stories them. Black religious scholar, Yvonne Chireau, points out that "the writing of history is about hearing from the dead, telling the story of the dead". What happens, I ask, following her questions, to storying place when "the dead speak back?" (quoted in Lemire (submitted). See also Chireau, 2003).

In re-membering, putting back together again, my memories spiral; they layer. This place I remember doesn't exist anymore, except it does. And the cemetery, it was and is full of stories, so many stories of lives we didn't know. Stories that were cut off. Stories that brought so many people, in death, to that place, to become that place, very literally as their bodies decomposed. Our connections, as children and as a fifty-something me writing now, were and are nourished by the cemetery's particular materiality, the trees and the headstones, the grass and the stones and the tokens left by families, by ghosts and memories, through layers of more-than-human stories. It was and is a human place, yet not only human, and full of a broader, different kind of humanity; mundane but also transcendental somehow, extra-ordinary, a place of death so embedded in life. This is a place that storied and stories me, and that I story now in turn. Re-membering. So many beings, histories and lives continue to call to me. Haunted and haunting.

References

Bawaka Country, Burarrwanga, L., Ganambarr, R., Ganambarr-Stubbs, M., Ganambarr, B., Maymuru, D., Wright, S., Suchet-Pearson, S., Lloyd, K., & Sweeney, J. (2016). Co-becoming time/s: Time/s-as-telling-as-time/s. In J. Thorpe, S. Rutherford, & L. A. Sandberg (Eds.), *Methodological challenges in nature-culture and environmental history research* (pp. 101–112). Routledge.

Bawaka Country, Suchet-Pearson, S., Wright, S., Lloyd, K., Tofa, M., Burarrwanga, L., & Maymuru, D. (2020). Bunbum ga dhä-yu t agum: to make it right again, to remake. *Social & Cultural Geography, 21*(7), 985–1001.

Best, A., & Ramírez, M. M. (2021). Urban specters. *Environment and Planning D: Society and Space, 39*(6), 1043–1054.

Bishop, M. (2023). Humility. Listen. Respect: three values underpinning Indigenous (environmental) education sovereignty. *Progress in Environmental Geography, 2*(3), 191–201.

Bishop, M., & Tynan, L. (2022). Finding perspective through our more-than-human kin. In K. Sims, N. Banks, S. Engel, P. Hodge, J. Makuwira, N. Nakamura, J. Rigg, A. Salamanca, & P. Yeophantong (Eds.), *The Routledge handbook of global development* (pp. 593–604). Routledge.

Butler, J. (2004). *Precarious life: the powers of mourning and violence*. Verso.

Cameron, E. (2008). Cultural geographies essay: indigenous spectrality and the politics of postcolonial ghost stories. *Cultural Geographies, 15*(3), 383–393.

Cavanagh, V., & Standley, P. (2020). Walking in the landscapes of our ancestors-indigenous perspectives critical in the teaching of geography. *Interaction, 48*(1), 14–16.

Chireau, Y. P. (2003). *Black magic: religion and the African American conjuring tradition.* University of California Press.

Daley, L., & Wright, S. (2022). Unsettling time (s): reconstituting the when of urban radical politics. *Political Geography, 98*, 102707.

Davies, D., & Rumble, H. (2012). *Natural burial*. Bloomsbury.

Davis, C. (2005). Hauntology, spectres and phantoms. *French Studies, 59*(3), 373–379.

Deering, A. (2012). *Over their dead bodies: a study of leisure and spatiality in cemeteries*. (Doctoral Dissertation,. University of Brighton.

Edensor, T. (2005). The ghosts of industrial ruins: ordering and disordering memory in excessive space. *Environment and Planning D., 23*(6), 829–849.

Edensor, T. (2008). Mundane hauntings: commuting through the phantasmagoric working-class spaces of Manchester, England. *cultural geographies, 15*(3), 313–333.

Francis, D., Kellaher, L. A., & Neophytou, G. (2005). The secret cemetery..

Gordon, A. F. (2008). *Ghostly matters: haunting and the sociological imagination.* University of Minnesota Press.

Gordon, A. F. (2011). Some thoughts on haunting and futurity. *Borderlands E-journal, 10*(2), 1–21.

Grande, S., & McCarty, T. L. (2018). Indigenous elsewheres: refusal and re-membering in education research, policy, and praxis. *International Journal of Qualitative Studies in Education, 31*(3), 165–167.

Haritaworn, J., Kuntsman, A., & Posocco, S. (2014). *Queer necropolitics.* Routledge.

Holloway, J., & Kneale, J. (2008). Locating haunting: a ghost-hunter's guide. *Cultural Geographies, 15*(3), 297–312.

Howitt, R. (2020). Unsettling the taken (for granted). *Progress in Human Geography, 44*(2), 193–215.

Jedan, C., Kmec, S., Kolnberger, T., Venbrux, E., & Westendorp, M. (2020). Co-creating ritual spaces and communities: an analysis of municipal cemetery Tongerseweg, Maastricht, 1812–2020. *Religions, 11*(9), 435.

Kwaymullina, A. (2020). *Living on stolen land*. Magabala Books.

Lemire, J. (submitted). *Black healing as relational and transformative: Processing, releasing and Reclaiming through Dance*. Submitted to Transactions of the Institute of British Geographers.

Lloyd, D. (2005). The indigent sublime: specters of Irish hunger. *Representations, 92*(1), 152–185.

Lorimer, H. (2019). Dear departed: writing the lifeworlds of place. *Transactions of the Institute of British Geographers, 44*(2), 331–345.

Luckhurst, R. (2002). The contemporary London Gothic and the limits of the 'spectral turn'. *Textual Practice, 16*(3), 527–546.

Maddrell, A., McNally, D., Beebeejaun, Y., McClymont, K., & Mathijssen, B. (2021). Intersections of (infra) structural violence and cultural inclusion: the geopolitics of minority cemeteries and crematoria provision. *Transactions of the Institute of British Geographers, 46*(3), 675–688.

Maddrell, A., & Sidaway, J. D. (Eds.). (2012). *Deathscapes: spaces for death, dying, mourning and remembrance*. Ashgate Publishing.

Marshall, U. B., Daley, L., Blacklock, F., & Wright, S. (2022). Re-membering weather relations: Urban environments in and as Country. *Urban Policy and Research, 40*(3), 223–235.

Mbembe, A. (2020). *Necropolitics*. Duke University Press.

Mcinherny, A. (forthcoming). *Ngarragi gudida: listening as Kin methodology*.

Moreton-Robinson, A. (2015). *The white possessive: property, power, and indigenous sovereignty*. University of Minnesota Press.

Okechukwu, A. (2022). Urban social hauntings: disappearing gravestone murals in gentrifying Brooklyn. *Environment and Planning D: Society and Space, 40*(1), 138–157.

Roberts, E. (2013). Geography and the visual image: a hauntological approach. *Progress in Human Geography, 37*(3), 386–402.

Rugg, J. J. (2022). Further remarks on modern sepulture: twenty years of cemetery studies and eight core questions defining cemetery research. *Eastern and Northern European Journal of Death Studies, 1*(1), 16–46.

Semple, S., & Brookes, S. (2020). *Necrogeography and necroscapes: living with the dead. World Archaeology, 52*(1), 1–15.

Sharpe, C. (2016). *In the wake: on blackness and being*. Duke University Press.

Shaw, W. S., Herman, R. D. K., & Dobbs, G. R. (2006). Encountering indigeneity: re-imagining and decolonizing geography. *Geografiska Annaler: Series B, Human Geography, 88*(3), 267–276.

Smith, A. S., Smith, N., Daley, L., Wright, S., & Hodge, P. (2021). Creation, destruction, and COVID: heeding the call of country, bringing things into balance. *Geographical Research, 59*(2), 160–168.

Smith, A. S., Smith, N., Wright, S., Hodge, P., & Daley, L. (2020). Yandaarra is living protocol. *Social & Cultural Geography, 21*(7), 940–961.

Stevens, V., & Tolbert, J. A. (2018). Beyond metaphorical spectrality: for new paranormal geographies. *New directions in Folklore, 16*(1), 27–57.

wa Thiong'o, N. (2009). *Something torn and new: an African renaissance*. BasicCivitas Books.

Watego, C. (2021). *Another day in the colony*. University of Queensland Press.

Woolf, V. (1922) [2003]. *A writer's diary: Being extracts from the diaries of Virginia Woolf*. London, UK: Mariner.

Wright, S. (2024). *Becoming weather: weather, embodiment and affect*. Taylor & Francis.

Yandaarra with Gumbaynggirr Country, Smith, A. S., Marshall, U. B., Smith, N., Wright, S., Daley, L., & Hodge, P. (2022). Ethics and consent in more-than-human research: Some considerations from/with/as Gumbaynggirr Country, Australia. *Transactions of the Institute of British Geographers, 47*(3), 709–724.

Yandaarra with Smith, A.S, Marshall, U. B., Smith, N., Murphy-May, L., Daley, L., Hodge, P., & Wright, S. (2024). What does country-led mean from who/where we are on Gumbaynggirr Ngambaa country? *Australian Geographer, 55*(4), 477–493.

5

Situating: Fading Polaroids of Place

Joseph Palis

Abstract Geraldine Pratt once said: "domestic space [can be a] a provocation, a refusal and reversal of conventional understandings of home as a bounded, protective space" (Pratt, 2004, p. 44). This chapter is the storying of a space of intimacy as multi-scalar and not tied to conventional notions of belongingness. It follows the narrative of a returning native who actively creates a place in an estranged home country by assembling materialities that correspond to a domestic space of an imagined home. Home-making becomes a process that transcends physical space and one that constantly makes, un-makes and remakes a liminal and intimate domestic space that connects the global and the national to the personal through the rearrangement of the material and the resistance to the physical.

Keywords Dislocation · Home · Domestic space · Belonging · Liminality · Alterity

U. Kothari et al., *Stories of Place*, https://doi.org/10.1007/978-981-95-4921-4_5

> [O]ne does not get lost but loses one's self, with the implication that it is a conscious choice, a chosen surrender, a psychic state achievable through geography. (Rebecca Solnit, *A Field Guide to Being Lost*, 2006, p. 6)

This chapter maps various personal stories of lost geographies. These stories come in the form of vignettes that locate tiny cartographies of dislocation. Some stories are difficult to tell and the way they are recalled can be murky. While some resonant feelings emanating from encounters with people and places stand out more than others, it has been a challenge to sift through which stories, previously shared with a group of intimates, can now be shared to a larger public. In deciding which story to share or not share is a consideration of what is relatable, accessible, jarring, disturbing and, like Proust's madeleines, can trigger a plethora of emotions, of a remembered past, or a bittersweet childhood memory. Or something else altogether.

Each of the three vignettes that follow are different stories of dislocation and of finding a way to create a world when the familiar falls away and the unfamiliar appears (Solnit, 2006). The epilogue reflects on these stories and my own realisation that the act of writing about place can spark sensations of discomfort and catharsis, and which stories can be told to a global audience of readers while remaining true to their intent. These fragmented vignettes of varying lengths are of moments captured by a subjective memory that evoke my entanglements with places, natures and emotions and reveal the possibilities of constructing a world where lived life can thrive and where connections to what matters in the present can be found.

* * *

The late Philippine writer, poet and activist Jose Maria Sison who lived in Utrecht as a political exile with his wife, once expressed his longing for his country. He wrote a poem called *Sometimes, the Heart Yearns for Mangoes* (Sison, 2013), where he described this longing for the familiarity of his native country. An excerpt from this poem reads:

> Far less than the home,
> And the flow of kith and kin,
> Unfamiliar and now familiar

Things and places trigger
The pain of sundered relations,
Of losses by delays and default.

Sison's liminal space is understood and keenly felt by exiles, refugees and those who have stayed for long in a foreign territory but who plan to return to the country of their birth. While it is possible to remain rooted in a given territory, there is a constant cloud of uncertainty, of being uprooted and 're-planted' someplace else. Edward Said has also spoken about the "unhealable rift between a human being and a native place, between the self and its true home: its essential sadness can never be surmounted" (Said, 2000, p. 173). It is this "crippling sorrow of estrangement" (Said, 2000, p. 173) that is continually the source of anguish for Said (and myself).

In Sison's poem, political circumstances prevent him from returning to the Philippines. As an exile and political refugee, his unrelenting fight against the 'exploiters of his people' for national liberation is a steady source of light that keeps his militant activism alive. A poem is a story told in a language that can be protracted, even incomplete, to convey feelings that flit in and cut and that can also be simultaneously experienced. A poem can powerfully communicate emotions that perpetually shape-shift. I liken Sison's poem to what Madge Clare (2014) calls 'a spark', a 'fulcrum' to think through his own melancholia for a land that remains out of his grasp. Sison continues:

Those who seek to rob the exile
Of home, kith and kin,
Of life, limb and liberty
Are the loudest to mock at him
Who is helplessly at sea,
Uprooted from his soil.

The well-purposed exile continues
To fight for his motherland
Against those who banished him,
The unwelcomed exploiters of his people,
And is certain that he is at home
In his own country and the world.

Fig. 5.1 Longing for mangoes is longing for a lost archipelago. *Source* "His Mangoes" courtesy of Dominique Amorsolo

Sison's poem is a story of his exile and his awareness of being *here* and *not there.* The cartographic aspect of his poem locates him in a territory he temporarily occupies but one that continually dislocates and alienates him. He identifies unnamed oppressors and exploiters who strengthen his resolve and fan the patriotic flames to keep 'fighting for his motherland'. It is the firm resolve to fight for national liberation that kept him at home 'in his own country and the world', the unequivocal commitment to advance the liberatory and emancipatory struggles of his people (Rosca, 2004) (Fig. 5.1).

Yet, it is Sison's yearning for mangoes that hit me. This was mentioned in passing in his poem:

> Sometimes, the heart yearns
> For mangoes where there are apples.

I remember the exquisite scent of near-ripe mangoes and the fleshy yellow insides of ripe ones. There is something in its taste that immediately reaches my heart like no other fruit can. Perhaps this is what Proust meant when he tasted madeleines? Mangoes remind me of my childhood when their abundance could not fully satiate our demand for them. When I was living abroad, I tasted mangoes bought from the supermarket but the tartness and sweetness did not fully come together. Eating them did not bring me 'home'. I would hanker for Philippine mangoes and I would think about a particular mango variety—*manggang kalabaw*—that tastes so exquisite. Yet when I returned to the archipelagic space of my birth, I would ironically *forget* to eat this fruit—the same fruit that I longed for and that haunted me when I was away. Perhaps it was homesickness that made me long for a mango: a desire to re-experience the humidity of the islands as a reminder of my connection to a putative hearthspace and to capture an uncapturable past. Although not an actual exile like Sison, the wound of my ruptured relationship to my homeland continued to fester after more than a decade of living away from my native country.

* * *

In 2015, I returned to the Philippines after living in North Carolina for more than a decade. I thought it was time to come home for good. I felt the need to serve the university, serve the country, and as some scholars (Canonigo, 2014; Sahay, 2014) point out: be there to reverse brain drain. With my newly minted PhD at that time, I am an instrument to the discipline of geography in the Philippines, and it provides me a chance to re-root myself to the country.

Going back to a place I know by heart was anything but soft and smooth—it resembled more of a crash landing. There are familiar sights, smells and sounds and yet despite the initial excitement, the aching familiarity I yearn for remains elusive or as Solnit again reminds us: the unfamiliar appears (Solnit, 2006). People in my neighbourhood and my colleagues in the university speak Filipino but I grapple with the complex ways different, unfamiliar meanings are now woven into conversations. It is strange to listen to a familiar language and not understand anything.

Like someone speaks the right words but in a different order. Ideas appear as though they are told backwards. Perhaps this is what reverse culture shock is all about. It is painful in tiny and invisible ways and I am always in a state of discomfort, not fitting in properly. When I was based in North Carolina, I would visit Manila during school breaks and would feel this mounting anticipation every time I was on a plane bound for Manila. But when this particular travel to return home, for good, happened, I was disoriented. I am back to an old and familiar landscape like the ones I associate with my childhood and linger in my memory. I had been keeping a picture in my mind of a world of sweet and cosy familiarity like what Sison yearned for. But these mental pictures are like polaroid photographs. After a while, their vivid colours and contours lose some detail and subsequently fade where only the lightest of outlines of people and places remain. The faded version now just evokes an idea of how things are and were. Can one be an alien to a native country? Can one be in a familiar space and feel strangely dislocated? Can one be homeless at home? How does one situate one's self without a map of meaning, a compass or a guidebook that re-orients one where the cardinal points of the heart used to be?

I remember one vivid dream I had: I was in a jeepney and I paid 20 pesos to the driver. He kept on driving and I said "*Sukli ko po sa beinte*" (My change from the 20 pesos, please), acknowledging that I know the fare was less than 20 and that he still needs to give me my change; but he replied "*Beinte po pag foreigner*" (I charge 20 pesos for foreigners). I woke up and remembered being upset about it and thought about it for a long time. I thought dreams are spaces of refuge where I get to inhabit a place I desire, even if momentarily. But in my dreamscape, I am still struggling to assert myself and convince people of my rootedness and of being at 'home'. What if the idea of home is just a mental picture, a faded polaroid? Can one be at home in the liminal spaces? In the in-betweenness?

* * *

Carving a space from the unfamiliarity of the familiar reminds me of a Stevie Wonder song: *Show me how to do like you /show me how to do it.*

Upon returning to the Philippines and resuming my job as a faculty member in geography, I would spend much time in my office on my 'home' campus. I would be there in the morning and return home past midnight to sleep and then leave again in the morning. This would go on for several months. I would perform the everyday routine of an academic. Then all of a sudden, the rainy season is back, a new president is sworn in, a small road disappears in our street and two years have passed by so quickly. I noticed a few visible grey hairs too when I looked in the mirror.

A friend from my Chapel Hill days visited me in Manila and remarked on how my office resembled the one I had in North Carolina State University where I taught for six years. The familiar pictures I adorned on my bookshelf, the type of calendars I like to hang around my office, even the wall clock—as though a replica of my living space while living abroad had been restaged. Was I unconsciously preserving the landscape of my former workplace, and re-creating it in the new place? It is a strange admixture of a place, a Kafkaesque landscape of confusion, an inner absurd.

My office is a time machine that brings me back to where I was before. It is like finding a reference point but, to quote from a diasporic film by Jonas Mekas, I was still 'lost lost lost'. Once again, I ask: is it really home or some made-up fantasyland where my family used to be. Most of them are gone now: father, mother, three siblings. Perhaps it was not finding the familiar folks who had peopled my earlier world that made me so lost in the present.

* * *

Just as a poem maps stories of exile, films can also be signifiers of cartographic anxieties experienced by those who are faced with the dilemma of situating themselves in permanent places. My doctoral research is on the postnational imaginary of the so-called national cinemas. Films interest me because I can 'get lost' in the diegetic space of the story and imagine lives that may not be within my own life world. I am fascinated by experimental, playful and subtitled cinemas that do not follow a Hollywood-prescribed genre. In 2006, while gathering data for my research project, I interviewed Filipino filmmaker and 'father of Philippine independent cinema', Kidlat Tahimik ('Silent Lightning') about his films and their cinematic articulations that favour a more local and, as he calls it, an 'indie-genius' sensibility. The stories in his films situate the country in a flux while remaining steeped in its myths and folklore. I will focus on the film that he became famous for: *Mababangong Bangungot,* and discuss another film—*Rigodon*—as a companion piece to problematise the elusive nature of being in-between places.

In 1977, Kidlat Tahimik completed his first film called *Mababangong Bangungot* (popularly known as 'Perfumed Nightmare' although I prefer writer E. San Juan's 'Fragrant Asphyxiations' translation). The film tells the story of a jeepney driver in a Philippine rural village named Kidlat who went to Europe to experience modernisation. Although initially excited, he became disillusioned when he witnessed the demolition and collapse of small markets to pave the way for "hideous concrete supermarkets ... [resembling] atomic power stations" (Jameson, 1992, p. 190). Kidlat's eventual alienation to manicured landscapes and rigid infrastructures in a sea of whiteness disillusioned him, which later led to his return to his country. One of the most striking symbolisms of the film is the presence of a village bridge where Kidlat crosses. If the bridge symbolises modernity and provincial life, Kidlat neither makes a decision to assign meanings nor to choose to assimilate in either side of the bridge. He treats them as autonomous units, or as Jameson worded it: "juxtaposed, side by side or in sequence" (Jameson, 1992, p. 192). This ambivalence supports Geraldine Pratt's view of the contradictory nature of identities. She finds that "we have multiple and sometimes contradictory subject positions and are sometimes torn between identifications" where situations, circumstances and places contribute to the "movement between identifications" (Pratt, 1998, p. 26). What struck me most about the scene of the bridge in Kidlat's film is how it brings up my own sense of (non)belonging as though I inhabit a liminal space. For me, places are unstable yet bounded and grounded to a territory. Whatever stability exists comes from movement, as though by moving in between them, a place becomes stable because of its permeability.

If *Mababangong Bangungot* is a parable about making a decision to live in another country permanently, or to stay in the homeland village,

Kidlat's action acknowledges transients who are in-between spaces and neither occupying nor making resolute decisions to occupy either side. To encompass exilic and diasporic filmmaking, Hamid Naficy proposes what he calls 'accented cinema': where films and the types of filmmaking document the "scattering of populations to places outside their homeland" (Naficy, 2001, p. 14). Accented cinema is characterised by the presence of "trauma, rupture and coercion" (p. 14). Because identities are fluid and not immutable, accented cinema is a celebration of the process of becoming that creates a dialogue between the homeland and what Naficy calls 'the hostland'. What resonates most with me is accented cinema's acknowledgement that there are transient lives that reside in the interstices of territories like Kidlat's character in *Mababangong Bangungot* and countless other cinematic stories of people in diaspora, who navigate spaces in foreign countries, and their various attempts to make the unfamiliar familiar. In an interview with Kidlat, he describes Filipinos as "sleeping typhoons locked in [our] colonial cocoons" (Palis, 2012, p. 16). Yet I must acknowledge that the transient experience can be problematic in light of the slippery definition of the term, its own attendant issues of inclusion and exclusion, and the specific discourses that govern various experiences and encounters with new environments.

Based on accented cinematic articulations, is it possible to have an unhomed cinema, of being 'homeless' at home, and not affiliated with an ontological anchor? In an earlier publication, I argued for an unhomed cinema that refuses a fixed residence even if the accented filmmaker is located in a territory that allows multiple perspectives of homelands, hostlands as well as ancillary and liminal spaces (Palis, 2008). Recognising migrant Filipinos who are occupying global spaces of labour, scholar Epifanio San Juan opines: "[w]e are … a quasi-wandering people, pilgrims or prospectors staking our lives and futures all over the world—in every nook and cranny of this seemingly godforsaken earth. Explorers and adventurers all" (San Juan, 2006, pp. 74–75).

If *Mababangong Bangungot* makes a case of transients occupying in-between spaces, *Rigodon* illustrates the tension between the archipelago and diaspora, and of the homeland and hostland. Released in 2005, *Rigodon* explores the lives of Filipino migrants in the post 9/11 United States. While firmly ensconced in the then still labelled 'First World', the ritual practices of the characters portrayed in *Rigodon* hark to village life in the Philippines that refuse easy categorisations of being *here* and *there*. Directed by the wife and husband team of Sari Dalena and Keth Sicat, *Rigodon* introduced three Filipino lives in New York: Amado, Salome and Dante. The character of Amado is an aging boxer who is desperately finding ways to legalise his stay in the United States by becoming (once again) involved in boxing—a sport he used to engage in, but now has to engage in again in order to stay legally in the country even if his physical might has long passed its peak. Salome is a mail-order bride and married to a US serviceman. She has openly articulated her desire to have a baby with blue eyes and possessing other desirable features that best capture and personify her American Dream. Yearning for assimilation and citizenship, Salome nevertheless performs distinctive and unmistakably Filipino practices such as wearing the traditional garb. Dante is an intellectual who philosophically engages with the idea of nationalism by imagining the Filipino community around him in New York as truly national, bringing none of the regional conflicts they may have in the Philippines. He has helped numerous Filipinos obtain legal status by circumventing certain immigration rules both as a form of resistance and diasporic nationalism against the country that both provided a space for his basic existence and his intellectual ruminations.

That the Filipinos in Rigodon characterise a transnationality is challenged by E. San Juan when he argues in his book *Filipinos Everywhere* that Filipinos "are not … transnationals" (San Juan, 2006, p. xiii). He reasons that "the putative 'Filipino' nation is in the process of construction, overseas Filipino contract workers have been considered transnationals or transmigrants—a paradoxical turn since the existence of the nation is problematic" (San Juan, 2006, pp. 42–43). If San Juan's assertion of the problematic nature of the Philippines as a nation is true, then it further enhances the unhomedness of Filipinos in the diaspora. It puts them in the interstitial space where homeland and their new environment do not even allow for a nation to be imagined.

The filmmaking practice and production ethic of *Rigodon* show the approaches privileged by the filmmakers. Dalena and Sicat grew up in the Philippines but situated *Rigodon* in the United States to narrate the lives of diasporic Filipinos against the backdrop of the aftermath of 9/

11. Dalena and Sicat said that their own experiences were similar to their adopted countries as they reflect on Filipino identities caught in constant transition. Unhomed cinema describes the interstitial space that Filipino films and their filmmakers occupy between homeland and diaspora. Whether the nation has a physical materiality or is imagined, unhomed cinema situates shifting Filipino identities against multi-ethnic praxis and multi-scalar mobilities. I find myself attracted to these fictional stories even if these are based in part on filmmakers' biographical backgrounds—they are my own too.

* * *

Epilogue

When I first wrote little vignettes of my experiences of, and connections to place, being dislocated and alienated was the dominant theme that appealed to me. I never thought about it at length but I wanted to write about it and the feelings and emotions I have about it. I wrote the original vignettes in Melbourne in the company of a storytelling collective who I have known for years. Perhaps it was my being abroad again, which led me to write continuously of my experiences with unbridled detail. When I revisited these stories a couple of years later, I was amused at the emotional outpourings that I allowed myself to write about my connection, or more specifically my disconnection, to places and the do-it-yourself wayfinding that I engaged in. I felt oddly disconnected with some of the stories that I wrote at that time. Some do not resonate with me now. What subjective frame was I in when I wrote them? If stories can be rewritten, what new avenues can open up in the process of storytelling?

Doreen Massey's notion of the 'simultaneity of stories-so-far' (Massey, 2005, p. 130) allows for the possibility that narratives are on-going, never fully complete, "a pincushion of a million stories".

Writing about being lost in a strange city, or any given territory, or not having emotional safety is a mixture of catharsis and apprehension. The writing and telling of stories of my connection to place by highlighting various geographies of lostness and occupying interstitial spaces allows me to acknowledge and make sense of the strangeness that comes from not resolving stories with neat endings, of stories where I found my emotional coordinates. If stories are maps, mine remain incomplete.

References

Canonigo, J. P. (2014). *Reversing the impact of brain drain in the Philippines (Blog)*. https://www.johnpaulcanonigo.com/2024/10/reversing-impact-of-brain-drain-in.html. July 9, 2025.

Clare, M. (2014). On the creative (re)turn to geography: Poetry, politics and passion. *Area, 46*(2), 178–185.

Jameson, F. (1992). *The geopolitical aesthetic*. University of Indiana Press.

Massey, D. (2005). *For space*. Sage.

Naficy, H. (2001). *An accented cinema: Exilic and diasporic filmmaking*. Princeton University Press.

Palis, J. (2008). *Cinema archipelago: A geography of film and the post-representational imaginary* (Unpublished dissertation). University of North Carolina-Chapel Hill.

Palis, J. (2012). Sleeping typhoons breaking out of their colonial cocoons: Conversations with Filipino filmmakers on cinema and nation. *Philippine Geographical Journal, 56*, 1–24.

Pratt, G. (1998). Grids of difference: Place and identity formation. In R. Fincher & J. Jacobs (Eds.), *Cities of difference* (pp. 26–48). The Guilford Press.

Pratt, G. (2004). Between homes: Displacement and belonging for second-generation Filipino-Canadian youths. *iBC Studies, 1140/i,* 41–68.

Rosca, N. (2004). *Jose Maria Sison: At home in the world—Portrait of a revolutionary (Conversations with Ninotchka Rosca)*. Open Hand Publishing House.

Sahay, A. (2014). Reverse brain drain: New strategies by developed and developing countries. In S. Sahoo & B. Pattanaik (Eds.), *Global diasporas and development*. Springer. https://doi.org/10.1007/978-81-322-1047-4_11

Said, E. W. (2000). *Reflections on exile and other essays.* Granta.

San Juan, E. (2006). *Filipinos everywhere: Displaced, transported overseas, moving on in the diaspora*. Ibon Books.

Sison, J. M. (2013). Sometimes, the heart yearns for mangoes. In *The Guerilla is like a poet* (pp. 140–142). Punctum Books/Uitgeverij. (Original work published 1994)

Solnit, R. (2006). *A field guide to getting lost*. Penguin Books.

6

Relating: Storying Atmosphere of Place

Maria Borovnik

Abstract Floating, bubbling, vibrant, turquoise, reviving, playful, blue, speedy, hopping, gorgeous, overgrown, muddy, green, shiny, soft, spluttering, cooling, healing, replenishing, life-giving, magically beautiful; a spritzy, energetic feel, a playful, laughing sensation, mossy smell, fresh taste—this is the atmosphere of the upper Te Awa o Te Atua. This story focuses on how the river and its landscape are actors interrelating with us. Interacting, connecting and communicating with nature, a walk along this river becomes a mix of visual, sensual, kinaesthetic, tactile and emotional experience. In this chapter, it is questioned whether the intermingling between self and nature is a purely organic matter that can be accessed by just anyone. Instead, questions of belonging and relating are tied in this discussion, highlighting the intrinsic relationality of tangata whenua genealogy that intertwines humans and more-than-humans.

Keywords Atmospheres · Sensations · Riverscapes · Non-human agency · Settler colonialism

U. Kothari et al., *Stories of Place*, https://doi.org/10.1007/978-981-95-4921-4_6

The original place names in Te Reo Māori are not presented in italics here, as is often found in academic writings. At this time, many original place names are still not normalised in maps across Aotearoa, and italicising perpetuates this othering. The English place names are therefore included in brackets following Te Reo Māori. Being most familiar with European terminology, I have used the term 'landscape' frequently in this chapter and I want to acknowledge Brian Murton (2013), who found there is more to the environment in an Indigenous context than the word 'landscape' could comprise. I am observing that Murton. (2013, pp. 149–150) has explained that naming the environment, or the equivalent of 'landscape' in te ao Māori (the world of Māori, who are Indigenous to Aotearoa), imprints the connection between the self (or body) and the environment, interlocking people, place and knowledge. In the original place naming process, an immediate connection between people and their ancestors was sealed. And therefore, the Indigenous place names allocated by tangata whenua (people of the land) are significant as both active relationships inwards and outwards between place and bodies are activated (Murton, 2013). The importance of these relationships must be respected by outsiders.

* * *

Unexpected and deep has been my connection to a river landscape with which I have no family ties. This unforeseen bond began right on the first day of the six days I spent at the upper Te Awa o Te Atua (Tarawera River) in the Waikato region of Aotearoa (New Zealand) during the summer transition of 2018–2019. This relationship seemed at first light-hearted, almost playful, a sensation of being in the 'now' without reasoning and without reflection. Then, in the process of finding words for this encounter, when writing my story, more and more awareness arose about this place as ara tawhito, a place of significance (Ruwhiu et al., 2022), filled with history, and whakapapa (genealogy), a place that needed to be remembered respectfully.

First, before entering the area, access permission had to be acquired from the local mana whenua (the authority of the land) at the nearest visitor information centre in Kawerau, which included a small financial contribution for each day in this area. Building our tent in the free camping ground, near Te Tapahoro (the Tarawera Outlet), my partner and I joined a mixed community of mostly local families and some travellers for our summer vacation. It is quite customary for many locals to go camping in the summer holidays. The camping space at Te Tapahoro was located on the Eastern side of Te Roto o Tarawera (Lake Tarawera) just at the corner of the river that sprung from it. A bridge crossed this transition between lake and river. Some chose to walk along the lake side, others the river. Kids would spend fun-filled days at the bridge jumping in and out of the river on one side, or the lake on the other. This was by far the most favourite location for kids and a safe playground where the camping community could keep an eye on their activities.

In the background, also watching, was *Te Maunga Tapu (Mount Tarawera), who looks like an ancient guardian, stern, earthy, and naked … attractive, almost hypnotising—demanding respect. This trapezoid volcano was part of the creation of the river landscape, over 11,000 years ago,* which was affected by a volcanic outbreak of Te Maunga Tapu, in 1886, resulting in a much younger fauna and flora (see Kawerau District Council, n.d.) (Fig. 6.1).

At the time of our summer at the river, we did not know these historical details. Instead, we joyfully observed the different moods and colourings of the river, which we perceived as 'innocent', not realising our own ignorance in not considering the underlying social and environmental histories and politics of this place. Then, my story started with:

> *Innocently, the river itself ripples away from the lake, then the reef, first muddy then becoming bluer and sometimes greener. It is not the algae overgrown type of green that has this stagnating, suffocating sensation in some rivers, but a vibrant, turquoise green of a hopping and floating light-heartedness. The stream is skipping and hovering over rocks and stones, sometimes thinner, sometimes deeper, always quite speedy.*

A friend, related to this region, told us about the campground and pointed out that a walk along the river would lead to the enormous Te

Fig. 6.1 Te Maunga Tapu is at the horizon of the upper Te Awa o te Atua. *Source* David Brown

Tatau a Hape (Tarawera Falls). Focused on finding these waterfalls we started our journey along the river and found ourselves drawn into its fascinating landscape, which seemed to become more and more mystical at every corner. We became so immersed with this river world that we did not reach the falls until our very last day: *We walked with the river, stopping at every corner, admiring the gorgeousness of the bush, its different colours with all the little details* of the river landscape, seen and unseen. The unseen was perceived—in a similar sense perhaps as Wylie (2006) described this as "immanent relation", where landscape merges with the one who views it. This process of unfolding of a landscape situates the world *within* a person (Deleuze, 1992 in Wylie, 2006, p. 531), rather than one where a person is externally placed. Whereas Ingold (2011, p. 115) finds that we are mingling with our surroundings, as we "breathe, the wind, light and moisture of the sky" and where these are bound and intermingled "with the substances of the earth", the sensations we had as we were walking extended to a more closely felt connectedness to the riverscape. Instead of just mingling with the river and its landscape, after a while it felt more like we all became one immersed world.

Along our walk, with *the stream skipping and hovering over rocks and stones, sometimes thinner, sometimes deeper, always quite speedy,* splattering along with us on one side, we noticed many different types of nature 'gatherings' forming along the walkway (Fig. 6.2):

> *If there are fairies and other nature beings, then they most certainly are here – somewhere. It is as though we can sense nature as actively alive. As though we could hear it speak to us. Or show us something magical.*

> *Each fern, even the littlest, vibrantly hanging out there, is having a good time. Moss in many different shapes is showing off its beauty. Whole clusters of thick, soft moss are calling out to be touched and laid upon. Looking up from one of these moss beddings, the much taller native bush leaves are interacting playfully with the blue sky. Following their dance, we can almost hear a melody humming sweetly. Just behind this scene, when we turn a corner, a mini desert landscape reveals gatherings of mānuka and kānuka. What's their conversations about?*

We stood in deep listening, and yet, could not understand their 'conversations' and rather remained in stillness, all our senses open to this enchanting world.

Landscape materialities are explained by Wylie (2006) as "the agency and autochthony of surfaces, reliefs, textures, and tones" (p. 533), which they may share with those open to noticing. More vividly expressed by Edensor, landscape "is alive with energies, eternally fluid, its rocks, earth, vegetation and climate continually changing as elements are entangled and folded together in a continual making" (Edensor, 2017, p. 617). It is the "myriad relationships" (Kanngieser et al., 2024) with these more-than-human beings that have kept us in awe and have had a strongly healing and transforming effect. The connection to and immersion with the river has brought us closer to our 'selves'.

> *Yet another bend on this path along the river and we are in awe: impressively massive, shiny, deep-black obsidian rocks are spluttered across our path. They*

Fig. 6.2 Nature conversing. *Source* David Brown

offer us to lean on, feel their shiny softness and their sharp edges. Without knowing how ancient these rocks really are, we feel like wanting to spend time with these tall, wise ancestors – and we are drawn back to them each day of our summertime in the area. Right behind, we find a hidden access to the river, offering a most wonderful swimming spot (*Fig.* 6.3).

Again and again, every day, we enter the river at this site, which looks like a turquoise paradise. We are surrendering into its currents. And each time it is as though we are entering a nature-spell, as though we can sense something more-than-usual playing with us, elating us in an exquisite spirit of laughter. We are becoming one with the river, wanting to be carried by it, dancing with it, floating away with it and back. Fresh and cool, healing and replenishing, the river eventually lets us sit on a tree branch, where we can hardly stop watching its movements, little and large, all of it filled with life. Water is life giving. The river is a song. The river is spellbindingly beautiful.

Authors who deeply engage with Indigenous thought and knowledge, such as Kanngieser et al. (2024), have prompted that relationality

Fig. 6.3 Obsidian Tree Embrace along Te Awa o Te Atua. *Source* David Brown

indeed requires respectful listening and an understanding that "more-than-humans come into being not as separate entities but relationally, as kin, with responsibility to each other" (similarly see Country et al., 2016; Hond-Flavell et al., 2021; Johnson & Larsen, 2013; Larsen & Johnson, 2017; Watene, 2022). Also, Larsen and Johnson (2017) have claimed that it "is through genealogical connections not only to ancestors, but also places, stories, and ideas that our reciprocal relationship with the world and life can be understood" (p. 179). These thoughts go further than those of writers fascinated with the atmospheres of places and their often affective interactions with air, light, darkness or weather (e.g. Bille et al., 2015; Collier, 2020; Cook & Edensor, 2014; Edensor, 2012, 2013,

2017; Edensor et al., 2020; Edensor & Lorimer, 2016; Ingold, 2010, 2011; Martin, 2011; Sørensen, 2015; Vannini & Bradley, 2020). In an Indigenous context, relational and reciprocative aspects take prominence. In most Indigenous contexts, places are so closely connected to self/body that mutual respect between kaitiakitanga (socio-environmental care takers) and place is part of relationality. Thus, authors who work closely within Indigenous settings urge to not (mis)use relational connections to these non-human elements as basis for academic analysis, but instead respectfully attend to the stories that are best understood by people who belong to the land (tangata whenua in Aotearoa, or Country in Australia) (for example, Kanngieser et al., 2024; Wright, 2020).

Some foundations of Māori values and ideas are, as Watene (2022) has explained, relations that are "protecting deep social, environmental and intergenerational connections or 'genealogies', enacting responsibilities in ways that nurture these relationships, and are cultivating collective transformation" (p. 19). Relational values include manaakitanga (practices of care and hospitality), kaitiakitanga, utu (reciprocity), and whakawhanaungatanga (relationship with others) (Larsen & Johnson, 2017), and also whai take (guidance to resource availability and use) (Ruwhiu et al., 2022). These practices and values involve fundamental responsibilities that include "people and other living things in ways that enrich our well-being" (Watene, 2022, p. 23). In Aotearoa, natural features, such as rivers, mountains, rocks or forest formations, are part of tribal affiliations, and may be ancestors; and these collective, genealogical connections between humans and more-than-humans involve responsibilities and exchange of care in meaningful ways. Larsen and Johnson (2017) even go as far to say that hospitality "emanates from place, and is granted through its authority and agency"; and then in return they explain that "it is our responsibility to show this hospitality through our actions" (p. 179). In my mind, such responsibilities must be respectfully acknowledged by everyone who approaches nature; whether we belong or not, it is our ethical duty to move with care for every being in the natural world. In Aotearoa, Te Awa o Te Atua and the accompanying Te Maunga Tapu will eventually have legal rights of personhood, much as Waikato Te Awa (the Waikato River) and Taranaki Maunga (Mount Taranaki) already have had recognised (Edirisinghe & Suchet-Pearson,

2023; Parliament of New Zealand, 2017; RNZ, 2005). With their strong relational roles, rivers and mountains in Aotearoa, as in Australia, must be respected as "more-than-human sovereignties" (Kanngieser et al., 2024).

Contemplating these relational connections in Aotearoa, while reminiscing about my time at the river and in the process of writing this chapter, I was painfully reminded of my positionality as a new migrant to Aotearoa. I arrived here in 1997. As tauiwi (foreigner), with a whakapapa from Germany and Slovenia, I recognise that I have no rightful relation to any of the places and landscapes in this country even though I have been deeply moved by some of these, and perhaps most significantly resonated with the Tarawera riverscape. Respectfully, I acknowledge that my storytelling is from an outsider perspective, and without rightful relation to the Tarawera region, my story and understanding lacks the deeper knowledge of these lands and the history of land and people. Despite my personal resonance with this riverscape: without this depth of belonging-by-birth and kinship, the full story of this place will not be adequately understood.

For several months brooding over my story, I was not sure whether it was possible to publish it. Consequently, crafting this chapter has opened a somewhat sorrowful journey of longing to belong, as my tauiwi positionality feels incomplete or even deficient. I sense the fragmentation of having left Cologne, my place of upbringing, and at the same time not being related to the land in Aotearoa I migrated to. I feel surprisingly uprooted and unsettled. During intensive korerorero (deep conversation) with tangata whenua colleagues, I understand that the respectfully narrated deep interaction that I had felt with the water, mosses, trees, rock and plant formations in this part of the river, could be explained (even published), by taking great care and caution, revealing a clear transparency of my non-Indigenous, tauiwi, perspective and the inherent lack of full knowledge and sound relationality within this positionality. It is hopeful to understand that in "Māori conceptions of landscape and place, emphasising that creation is a process of continuous action that comes about through the unfolding of Te Kore, the primal power of the cosmos that already contains the form of every possible being" (Murton, 2013, p. 146), and where "cadences of ancient songs, of ritual calls, of sacred chants through which the world is sung into existence … and the relationships are sung which bind all together within the cosmos" (p. 147), may potentially breathe "life into the network of subtle interconnections between human beings and the entire natural world" (Stewart-Harawira, 2005, p. 18, in Wright, 2020, p. 210). My hope is that the unfolding of place itself in this world of emergence might accept this tauiwi, who is witnessing in awe.

> *We were so grateful to have spent time at this river, Te Awa o Te Atua, which was calling us back every day, during our week-long summer holiday. We bathed in gratitude for the strong replenishing gift that the river continued to present to us. Each day we had the urge to walk along the fairy landscape, greet the obsidian ancestors, meet with the same bubbly blue-greenish magnificence and connect with it. We felt like wanting to be part of this life-giving nature, feel it on our skin, and be in direct contact with it. Not just with the water but with all the nature beings this charming landscape had to offer.*

It is now that I am re-reading non-Indigenous academic writings, such as Ingold's, which I have relished in the past. Ingold (2011) has engaged with landscape as part of a larger world, with 'solid' land that is connected to 'empty' air, in which light, wind and weather operate. It is in this open, and yet, bounded, world, in which bodies move, dwell, or intermingle. In Ingold's view, there are different layers to landscapes, where

> the 'ground' is not really a coherent surface at all but a more or less impenetrable mass of tangled undergrowth, leaf litter and detritus, mosses and lichens, stones and boulders, split by cracks and crevasses, threaded by tree roots, and interspersed with swamps and marshes overgrown with rafts of vegetation that are liable to give way underfoot. (p. 119)

The walking experience is then placed in between the ground and "the clear sky above" (p. 120). Earlier, Ingold (2010) expanded on ideas that the activity of walking could be seen as knowledge making: while we create paths and tracks, we 'make' our way in this world (p. S122). This active intermingling and 'making way' is in contrast to Murton's (2013) deliberations on te ao Māori, where knowledge is already imbued within the environment, and is interacted with in the process of naming places and in remembering the knowledge that these names entail. Yet, Ingold (2011) also acknowledges that the "ground appears to be bursting with life and movement" (p. 121). He relates this active ground, in which substances intermingle, with a life-force "which will not be confined within bounded forms but rather threads its way through the world among the myriad of lines of its relations" (p. 124). In his writings, it is possible to relate to a sensual understanding of an environment in which one is moving and mingling. There is, however, something missing that other authors have emphasised within an Indigenous context, which is the strong *reciprocal care* between human and non-human worlds (as described by Larsen & Johnson, 2017; Ruwhiu et al., 2022; Watene, 2022). Knowledge gained in a European context comes across as limited, in that the deeper connection to ancestry in nature is absent; and the focus is on tangible materialities, such as the ground below my feet, as Ingold (2010) suggests. And yet, his realisation that materials are in motion within a "world without objects" might offer the possibility of a certain transcendence, of merging with "a world in flux and becoming" (Ingold, 2011, p. 131), which is aspiring.

At the start of this chapter, I explained that engaging with landscape as an 'innocent' interactor would not be enough to understand the deeper issues within places. In this context, Edensor (2017) has shown that a landscape has underlying power relations that are imprinted in regulations, as can be found in many European contexts, shaped by symbolic significations and "the perpetuation of dominant ideas about *who* belongs in landscape" and "what pursuits are preferable" (p. 596); places in these settings, as Edensor et al. (2020) explained, "continuously (re)emerge out of durable and resonant historical formations" (p. 15). No wonder that authors, such as Ruwhiu et al. (2022), find that Western perspectives do not adequately understand Indigenous world views and links between identity and landscape. Landscape in Europe, indeed, is structured in political, measurable entities. These hegemonic contexts were prescribed to colonial settings without considering the intertwined relations between place and people, that nourish, heal, guide and teach in their "more-than-human entanglements" (Wright, 2020, p. 209). In these entanglements, place cannot easily be talked about. Sarah Wright explained:

> Coming into being with place is done with place, not with words, and in ways underpinned by diverse cosmologies and Indigenous Laws/lores and, crucially, in ways embedded with differential relations of power, positionality and sovereignty. (p. 210; my emphasis)

In other words, relationality always includes issues of power and sovereignty, where it would be crucial to understand how and where tino rangatiratanga (self-determined sovereignty) is practised, and where colonial misrepresentations of people and places continue. Larsen and Johnson (2017) suggest that the hospitality some places offer cannot be seen as entitlement, or taken-for-granted by visitors. Instead, there are boundaries to not continue extractive processes of dispossession. Places, however, offer opportunities as "through our relationship with place … human and nonhuman understandings of the universe are possible (Larsen & Johnson, 2017, p. 179), and these understandings open up "uncomfortable moments in which our taken-for-granted assumptions about the world begin to crack" (p. 181, referring to Seawright, 2014, p. 558). These 'frictions' are hopeful, as they may lead to insights about our positions within these relationships. If we honestly recognise our own contributions to structures of privilege, we might begin to intentionally turn these around, starting with the way we communicate with place. Interacting with places as ancestors, knowledgeable, sentient beings respectfully includes acting and expressing gratitude for the beauty we receive, and looking for possibilities to replenish where harm was done.

References

Country, B., Wright, S., Suchet-Pearson, S., Lloyd, K., Burarrwanga, L., Ganambarr, R., & Sweeney, J. (2016). Co-becoming Bawaka: Towards a relational understanding of place/space. *Progress in Human Geography, 40*(4), 455–475.

Bille, M., Bjerregaard, P., & Sørensen, T. F. (2015). Staging atmospheres: Materiality, culture, and the texture of the in-between. *Emotion, Space and Society, 15*, 31–38.

Collier, M. (2020). Place-walking. The Umwelt explored through our creative imagination. In T. Edensor, A. Kalandides, & U. Kothari (Eds.), *The Routledge handbook of place* (pp. 678–691). Routledge.

Cook, M., & Edensor, T. (2014). Cycling through dark space: Apprehending landscape otherwise. *Mobilities, 12*(1), 1–19.

Deleuze, G. (1992). *The fold: Leibniz and the Baroque*. University of Minnesota Press.

Edensor, T. (2012). Illuminated atmospheres: Anticipating and reproducing the flow of affective experience in Blackpool. *Environment and Planning D: Society and Space, 30*, 1103–1112.

Edensor, T. (2013). Reconnecting with darkness: Gloomy landscapes, lightless places. *Social and Cultural Geography, 14*(4), 446–465.

Edensor, T. (2017). Seeing with light and landscape: A walk around Stanton Moor. *Landscape Research, 42*(6), 616–633.

Edensor, T., Kalandides, A., & Kothari, U. (2020). Introduction. Thinking about place—Themes and emergent approaches. In T. Edensor, A. Kalandides, & U. Kothari (Eds.), *The Routledge handbook of place* (pp. 1–11). Routledge.

Edensor, T., & Lorimer, H. (2016). "landscapism" at the speed of light: darkness and illumination in motion. *Geograpfiska Annaler: Series B, Human Geography, 97*(1), 1–16.

Edirisinghe, A., & Suchet-Pearson, S. (2023). Nature as sentient being: Can rivers be legal persons? *RECIEL: Review of European, Comparative and International Environmental Law, 33*, 224–235.

Hond-Flavell, E., Theodore, R., Treharne, G. J., Tamati, A., Edwards, W., Poulton, R., Hond, R., & Ratima, M. (2021). Tangi te Kawekaweā. Whānau engagement in Kaupapa Māori early years provision—An exploratory qualitative study. *MAI Journal. A New Zealand Journal of Indigenous Scholarship, 10*(1), 3–16.

Ingold, T. (2010). Footprints through the weather-world: Walking, breathing, knowing. *Journal of the Royal Anthropological Institute, 16*(s1), S121–S139.

Ingold, T. (2011). *Being alive: Essays on movement, knowledge and description*. Routledge.

Johnson, J., & Larsen, S. (2013). *A deeper sense of place. Stories and journeys of collaboration in Indigenous Research*. Oregon State University Press.

Kanngieser, A. M., Soares, F., Rubis, J., Sullivan, C. T., Graham, M., Williams, M., Palis, J., Tynan, L., Daley, L., Blacklock, F., Greenhough, B., Suchet-Pearson, S., Wright, S., Lloyd, K., & Marshall, U. B. (2024). Listening to place, practicing relationality: Embodying six emergent protocols for collaborative relational geographies. *Emotion, Space and Society, 50*, 101000.

Kawerau District Council. (n.d.). *The history of Te Tatau a Hape/Tarawera Falls and Te Tapahoro/Trawera Outlet.* https://www.kaweraunz.com/explore/leisure/history-te-tatau-hapetarawera-falls-and-te-tapahorotarawera-outlet [Accessed 26 June 2025].

Larsen, S., & Johnson, J. (2017). *Being together in place: Indigenous coexistence in a more than human world*. University of Minnesota Press.

Martin, C. (2011). Fog-bound: Aerial space and the elemental entanglements of body-*with*-world. *Society and Space, 29*, 454–468.

Murton, B. (2013). From landscape to whenua: Thoughts on interweaving Indigenous and Western ideas about landscape. In J. Johnson & S. Larsen (Eds.), *A deeper sense of place: Stories and journeys of collaboration in Indigenous research* (pp. 139–156). Oregon State University Press.

Parliament of New Zealand. (2017). *Te Awa Tupuna (Whanganui river Claims Settlement) Act 2017*. http://www.legislation.govt.nz/act/public/2017/0007/17.0/DLM6830851.html

Radio New Zealand (RNZ). (2005, 31 January). *Taranaki Maunga becomes a legal person as treaty settlement passes into law.* https://www.rnz.co.nz/news/national/540431/taranaki-maunga-becomes-a-legal-person-as-treaty-settlement-passes-into-law (Accessed 11 July 2025).

Ruwhiu, D., Amoamao, M., Carter, L., Bargh, M., Ruckstuhl, K., Carr, A., & Awatere, S. (2022). Ngā Whai Take. Reframing Indigenous development. In K. Ruckstuhl, I. A. Velásquez Nimatuj, J. A. McNeish, & N. Postero (Eds.), *The Routledge handbook of Indigenous development* (pp. 297–308). Routledge.

Seawright, G. (2014). Settler traditions of place: Making explicit the epistemological legacy of white supremacy and settler colonialism for place-based education. *Educational Studies, 50*(6), 554–572.

Sørensen, T. F. (2015). More than a feeling: Towards an archaeology of atmosphere. *Emotion, Space and Society, 15*, 64–73.

Stewart-Harawira, M. (2005). *The new imperial order: Indigenous responses to globalization.* Zed Books.

Vannini, P., & Bradley, A. (2020). Weather and place. In T. Edensor, A. Kalandides, & U. Kothari (Eds.), *The Routledge handbook of place* (pp. 135–144). Routledge.

Watene, K. (2022). Indigenous development as flourishing intergenerational relationships. In K. Ruckstuhl, I. A. Velásquez Nimatuj, J.-A. McNeish, & N. Postero (Eds.), *The Routledge handbook of Indigenous development* (pp. 19–27). Routledge.

Wright, S. (2020). Thinking, doing and being decolonisation in, with and as place. In T. Edensor, A. Kalandides, & U. Kothari (Eds.), *The Routledge handbook of place* (pp. 208–217). Routledge.

Wylie, J. (2006). Depths and folds: On landscape and the gazing subject. *Environment and Planning D: Society and Space, 24*, 519–535.

7

Journeying: Geographies of Belonging in a Fractured World

Karen Paiva Henrique

Abstract This chapter attends to the ways in which storying place can help us locate and navigate our positioning in the world. By storying a university bursary—a seemingly mundane bureaucratic building—I trace my own journey as an international scholar, from earlier movements in life to building an academic career across Brazil, the United States, Australia and the Netherlands. In storying my journey through the bursary-place, I challenge assumptions about what places, and thus what journeys, are worthy of being storied and what countering persistent silences can mean for those who continue to feel and remain marginalised, also within academic circles. Ultimately, the chapter foregrounds the entanglements of places that exist at the intersections of multiple life stories and a sense of belonging, while staying attentive to the uneven political possibilities of being and becoming through place in a fractured world.

Keywords Everyday · Embodiment · Liminality · Borders · Identity

This chapter centres on what counts as place in storying our journeys of being, becoming and belonging in the world. As I sit down in front

U. Kothari et al., *Stories of Place*, https://doi.org/10.1007/978-981-95-4921-4_7

of the computer to write this chapter, I feel a sense of paralysis. For the first time in a very long time, I am unable to overcome the blank page staring at me. Writing this chapter should not be such a difficult task. After all, it has been more than a year since I first wrote the short story about a place that I would share with our storying collective and that would become the genesis of this chapter. We had many conversations about our stories of place, their meaning to us, what they (can) evoke in others, and the relations that bring them together and keep them apart. I have had ample opportunity for self-reflection. So, I am not sure how or why this has become such a daunting task.

Not only that, but place and storying have long been central to my work as a critical climate adaptation scholar. For over a decade now, I have sought to make visible how climate change comes to exist to people in their everyday lives and to highlight how climate-induced impacts are unevenly experienced and addressed in and through the places people inhabit, places that have profound meaning to them (Henrique et al., 2022; Henrique & Tschakert, 2019, 2022). I develop my work in conversation with other critical geographers and adaptation scholars (e.g. see Ajibade, 2017; Alvarez & Cardenas, 2019; Betteridge & Webber, 2019; Colven, 2017; Goh, 2019; Padawangi, 2014) who call attention to the perverse outcomes of adaptation policies and practices predicated on the displacement and dispossession of historically marginalised groups and who advocate for situated approaches that allow people to adapt in place on their own terms.

Crucially, my work builds on contributions from Latin American and feminist scholars to demonstrate how the production of place allows individuals and communities pushed to the margins of urban development to reclaim spaces they have been systematically excluded from (Brown & Knopp, 2003; hooks, bell., 1984; Manzi et al., 2018; Porto-Gonçalves, 2017). This work is attentive to the entanglements of body-space-subjectivities, as individuals mobilise the (limited) resources they possess to build and maintain communities worth living and fighting for—an effort exerted through their bodies through 'sweat, sore muscles, calluses and bruises' across years of dedicated physical and emotional labour (Henrique, 2023; Truelove, 2011). Ultimately, this work foregrounds how individual and collective efforts to produce place translate into a sense of belonging, place attachment and place-based political subjectivities with the potential to transform sociopolitical relations that guide climate adaptation in and beyond the contours of the bounded geographical spaces in which they emerge (Henrique, 2023). In short, place and storying place as meaning-making practice are central to how I think about and articulate transformational political possibilities in relation to climate adaptation and beyond.

Yet, when trying to story a place of my own, I pause.

As I try to unpack the meanings of the place I have chosen to story—an ostensibly mundane university bursary that was designed to handle the institution's financial flows—I am puzzled by my own choice. The bursary is certainly not the most evocative of places. It is essentially a place of passage; students only visit the bursary to manage their financial affairs and occasionally access academic records. It is a place where it may be assumed that no meaningful relations are forged between visitors, and between visitors and place, unless of course one happens to work there. In this sense, the bursary could qualify as Marc Augé's (Augé, 1995) notion of a 'non-place'—i.e. as non-relational, a-historical and unconcerned with one's identity. Yet, I find it too easy to dismiss the bursary as a non-place, one that is not worthy of being storied.

So, I decided to embrace this pause as an opportunity to take the meaning of the bursary as place in my own personal and academic journey seriously. Here, taking such an effort seriously means questioning academic assumptions about what places, and thus what lives, are worthy of being storied; in which ways these assumptions become internalised by academics themselves, and the absences and silences they produce within institutional life; and what countering these silences can mean for those who continue to feel and remain marginalised within academic circles. I do so while also reflecting on the political possibilities addressing these questions can offer for decentring and democratising epistemologies and ontologies within academia. To do so, in what follows I story my own journey in and beyond the academy as I endeavoured to find my place (i.e. to fit in and belong) in different academic institutions across the world, and the challenges and emergent sense of liberation I have encountered in doing so.

In searching for a beginning to my story of place, I am inspired by a good friend and colleague at the University of Amsterdam, Maggi Leung, a migration and mobilities scholar whose work centres on tracing the trajectories of those who choose, or are forced, to move. In her autobiographical account of mobility and internationalisation in higher education, Leung (2025) takes us (the readers) to the very beginning of her education trajectory, explaining how it is deeply entangled with that of her family and their own mobility pasts. Also a first-generation scholar, Leung reflects on how her concerted efforts to question and counter hegemonic thinking cannot be disassociated from her non-elite, working-class and international background. In her account, I find the beginning of my own story.

* * *

Being

My mother held a deep passion for education. She was a top student who achieved the highest marks in her class. She also came from a very modest background, which meant that when growing up, as the eldest daughter, she was entrusted with taking care of her three siblings while my grandparents worked hard trying to make ends meet.

While my grandfather moved between different, generally low-paying jobs (I remember my grandfather being away for long stretches of time when he was a bus and truck driver), my grandmother tried to supplement the family's income by selling cosmetics door-to-door. We always joked that my grandmother could sell anything to anyone, a fact that was confirmed by the many awards she won over the years for being a top-seller (a collection of porcelain dolls she proudly displayed in a wall-mounted shelf in her house). It was when helping my grandma with her deliveries that my mom would meet my dad. They were soon married, starting a family of their own.

Growing up, our family moved a lot, at least more than any other family I knew. When I was born, my family lived in Santana do Livramento in the South of Brazil, at the border with Uruguay. My father, who was in the army at the time, was stationed there for a couple of years. He could not, however, 'move up the ranks' without having a university degree and we eventually moved elsewhere so he could find a better-paying job. I was two when we moved to Santa Cruz do Sul. A few years later we would move again, now to Novo Hamburgo. Both towns were former German immigration centres in the South of Brazil. Eventually, my sister and I moved to the state capital, Porto Alegre, to attend university. My parents soon followed.

Part of this mobile life trajectory was the fact that my father hardly ever worked in the same city we lived in. For most of my childhood, my father worked over an hour away from home. He would drive to work very early in the morning and come back late at night. When I was older and he had to move jobs, his commute became much longer; one and a half hours by plane. He would leave in the morning on Monday and only come back on Friday, at the end of the day. Yet, despite his physical absence from our home, my father was a very present father. He knew every detail of our lives.

My mom, my two sisters and I would occasionally travel for the summer holidays. My father stayed behind for work. At the end of every day, we would walk to a payphone nearby so we could talk with him. These were very brief calls, but long enough for him to ask each one of us about our days. When I moved away from Brazil, we continued this tradition. We called each other every Sunday and I realised how much attention he paid to my life, always asking for more details. Maybe the fact that he had to commute so much when we were young and had to stay away during the summer to ensure he could provide us with a good life helped normalise the idea that you can still nurture your family from far away. It probably also normalised a sense of uprootedness that I carry with me to this day.

In Portuguese, when we want to say someone is *from* a certain place, we use the preposition *de*, which has a double meaning. It means that a person is *from* that place, having been born there. But it can also mean that a person is *of* that place, that they *belong* to that place. Growing up,

I never felt I was *de algum lugar específico* (or, from any specific place). I do not remember feeling particularly attached to any of the places where we lived.

I also remember in vivid detail how much my mom felt like we never truly belonged to any of those places. Not because she did not want or try to belong. On the contrary, she always tried very hard to make space for us. Our neighbours however were never welcoming. I remember thinking about what it was that made us so different (and undesirable) in all those different places. As other critical geographers recount in relation to their own life trajectories (Kinkaid et al., 2022; Leung, 2025), it was within the discipline that I finally gained the vocabulary to work through and articulate the systemic exclusions my mom experienced across her life, as well as the exclusions I encountered every time I moved to a new (academic) place.

* * *

Becoming

It was August 2005. I had left the cold winter of southern Brazil and travelled across hemispheres for a semester-long exchange programme at The Pennsylvania State University in the United States. This was my first time studying abroad, and my first time in a US campus, which, until then, I had only seen in Hollywood movies.

I remember approaching this large building, in the farthest northeast corner of the university campus; its prominent brick facade matching the style of the surrounding structures—and the campus planning guidelines, whose specific and homogenising details I would later learn. There were two perpendicular flights of stairs leading to the main entrance. Inside, a large hall welcomed visitors, surrounded by floor-to-ceiling windows partitioned in several small glass panes. Blue carpets resembling the school colours lined the entire floor and together with tall, eggshell-painted walls up to a high-ceiling, created a somewhat comforting space.

Two parallel rows of desks demarcated a long corridor where a group of students lined up, sporting a mix of university gear—caps, t-shirts and sweaters—all of course bearing the school's logo. As a group of students were called to different desks, one by one, another group would enter the room, swiftly taking their place in the queue.

There, I waited for my turn, taking it all in. The immensity of that space. The immensity of that moment.

Almost 20 years have passed but I remember entering this building as it had happened just a few moments ago. And although I seem to recall this space in detail, what I remember the most is its smell—the smell of old carpets, busy student life and bureaucracy—a new yet strangely familiar smell that I will forever associate with that space. That I will always associate with that moment.

This was the bursar's office known as the Shields Building (probably named after one of Penn State's many alumni donors), an administrative building that I visited perhaps 4 or 5 times over the nearly 5 years that I lived in State College—first as an undergraduate student and then for my master's and the first year of my PhD. But that smell stayed with me. Until now, every time I sense it, I am immediately transported back into that space, back into that moment, and that brings me joy.

Not because that was a particularly special building. On the contrary, I cannot think of a more mundane building in Penn State's enormous infrastructural apparatus. Remembering that building and that moment brings me joy because it takes me back to a very important moment of my life. That was the first time I felt I was part of an international academic community, where I found myself connected, part of something bigger. Bigger than Brazil, bigger than the United States.

For a long time, I wanted to study abroad. I imagined how life would be in other places, learning in other languages, learning from different cultures. Maybe it all started when my mom was offered a full scholarship to study in the United Kingdom, but my grandfather did not allow her to go. I think it was she who instilled in me this drive that would take me to many foreign lands.

What is interesting is that, many years later, when I arrived at the University of Western Australia for my PhD (and later my Postdoc) and entered the geography building, I sensed that same smell from the bursar's office at Penn State. All those feelings came rushing back. I was reminded I was a part of something bigger. This was a totally different building, much older, made of different materials, but the smell was the same. The feelings were the same.

Reflecting on the connections between these two utterly different places and myself in relation to them, I realise that I do not associate that smell with the bursar's office. I associate that smell with looking for my place in the world.

* * *

Belonging

In critically positioning my decision to story the university bursary in relation to my broader personal and academic journey and its many associated geographies, I can start to make sense of the importance such an ordinary place had for me. In that reading, the bursary becomes a gateway or a threshold for broadening the reach of my being and becoming in the world. In her poem "Intersection", Koleka Putuma (2024, pp. 28–29) eloquently speaks of a place in-between in which difference meets and can be reworked as freedom. Her words invite thinking about the bursary as *intersection* where life trajectories cross, the aspirations of higher education (for many still only a dream) are formalised or upended, and pathways towards a different future are forged, sometimes in unexpected ways. For me, the bursary symbolises gaining the possibility and permission to enter an 'elite' academic space. A possibility that was only attained through the many trade-offs made by my parents, including trading off the ability of being rooted in a specific place.

As intersection, the bursary becomes *place* through a set of relations akin to the flows and interconnections that underline Massey's (2013) call to reframe place in geography:

> ...what gives a place its specificity is not some long internalized history but the fact that it is constructed out of a particular constellation of social relations meeting and weaving together at a particular locus. (p. 155)

Read as intersection, the bursary-place challenges neat binary assumptions that define a sense of place, an attachment to place, and a sense of belonging as fundamentally irreconcilable with places that are not inhabited or occupied in expected ways, or through extended time. In some ways, the bursary-place becomes a liminal space 'betwixt and between' (Lugones, 2003)—not fully or properly either place or non-place—emerging as space with transitional potential for a "...metaphorical crossing of some imagined or temporal threshold" (Pritchard & Morgan, 2006, p. 764; cited in Parikh, 2019b, p. 3). Situated as liminal space, as intersection, the bursary-place offers "a more nuanced understanding of transformation, emergence and becoming" (McConnell, 2017, p. 142). In this way, the bursary-place calls for a reflection of how storying seemingly ordinary and unaccounted for places can open space for a prefigurative politics for decentring and democratising the production of knowledge, and its associated places. It also allows me to reposition myself as *of* and *from* the intersection (or *da interseção*)—where lives cross, are redefined and transformed through encounters with other people travelling their own unique and situated journeys through place.

It has been many years since I left State College. Since then, the possibility and hope I had of becoming with and belonging in an international academic space has been recurrently and stubbornly simultaneously remade and undone. As I joined and learned to navigate different institutions of higher education in different countries (the United States, Australia and the Netherlands), I continue the work my mom has started many years ago, that of making space for myself.

I have learned to situate myself in relation to systems that insist on 'othering' and excluding scholars from elsewhere (both students and staff), particularly those from so-called 'developing' geographies. Although the framing might have changed (we might now be referred to more often as scholars from the global South), the positioning is still one of otherness. Such othering happens not only in structural ways, through uneven patterns of valuation and recognition, but also in equally insidious everyday interactions. Othering in academia becomes manifest, for example, in the subjects and places we are allowed to study (reinforced by institutionalised funding mechanisms and the invitations we receive to present our work); the ways in which our positioning is compartmentalised and siloed to be drawn on and enjoyed only when strategic and beneficial for those keeping the gates of epistemic relevance; and in everyday interactions as we are measured for both our academic and non-academic interests and experiences. As Kinkaid et al. (2022) aptly explain:

> [s]uch incidents are repetitive, exhausting scholars who already labor disproportionately to establish the worth of ourselves and our scholarship in the academy. (p. 1.567)

It is against this backdrop that I reflect on my initial paralysis of storying the bursary as place and in relation to my personal and academic journey, as I also have been met with indifference, sidelining and devaluing when trying to relate my personal stories and experiences in academic settings.

Yet, at the intersection (here tentatively captured through the bursary-place, while acknowledging all the exclusions it (re)produces) one also finds alliances, kinship and hope. At the many intersections I have crossed and inhabited in my international academic journey, I found others with whom to share personal stories of and through place, and to challenge this discomfort. In one of our many conversations about our lives away from Brazil, my partner asked me: 'Are you ok with forever being the outsider, with never truly being accepted as the same?' Staying with the discomfort of being the 'other' in academia and otherwise, rather than brushing it aside, has helped me in the continuous work of establishing roots and becoming emplaced while accepting the reality of never fully belonging. My answer to his question was decisive: 'yes'.

* * *

Beyond

Attentiveness to my positioning in the bursary-place as it is defined in relation to my family's trajectory, their choices and my own experiences and choices also calls attention to the silences and absences that are constituted in and through place. The very existence of the bursary as an office whose mandate is to guarantee the maintenance of university money flows calls attention to a highly neoliberal knowledge landscape that is still unattainable for most. As Massey (2013) posits:

> ... it does seem that mobility and control over mobility both reflects and reinforces power. It is not simply a question of unequal distribution, that some people move more than others, and that some have more control than others. It is that the mobility and control of some groups can actively weaken other people. (p. 150)

A critical, situated and historical reading of the bursary-place also calls attention to those who have been entirely erased from this landscape of knowledge production (both epistemologically and ontologically), including the rightful owners of the land upon which Penn State campuses have been built—the Erie, Haudenosaunee (Seneca, Cayuga, Onondaga, Oneida, Mohawk and Tuscarora), Lenape (Delaware Nation, Delaware Tribe, Stockbridge-Munsee), Monongahela, Shawnee (Absentee, Eastern and Oklahoma), Susquehannock, and Wahzhazhe (Osage) Nations (Novotny, 2025). Such erasures are also ingrained and reproduced in and through the geography building at the University of Western Australia (the second place narrated in my story), which was built on Whadjuk Noongar Country. Their untold stories perpetuate and deepen these erasures with serious implications for which worlds are made possible or not.

Working through the question of what counts as a place worthy of being storied and positioning the bursary-place as intersection calls attention to the affordances of thinking with and storying through places that evade clear categorisations. This means embracing not only ordinary places and the (seemingly unimportant) relations that unfold within them as key to understanding geographies of difference but also positioning them as sites of intersection of multiple life stories. In my work, such recognition invites an explicit engagement with places, people and relations taken for granted also in climate research.

Moreover, and related, this is an invitation to think through how these intersections, and individuals' uneven positioning within them, emerge and are constitutive of relations within the very spaces of knowledge production. This is a call for critical geographers, among other academics, to account for the many stories, journeys and places that are lived through and have meaning for those who are still othered within the discipline and beyond, and to embrace the possibilities such pluralisation may afford to value, recognition and belonging in academic settings.

Finally, this is also an invitation to those who are situated at the margins of academia and allies to meet at the intersections and "[re]claim our experience as a source of critical knowledge" (Kinkaid et al., 2022, p. 1159; see also Parikh, 2019a) towards another academy, one that is plural, relational and dynamic. As Pierre et al. (2020) insist: "[t]o decolonise the methods of knowing and being requires us to reform, even revolutionize, the relationships among multiply situated knowledges and knowers in incommensurable worlds and journeys" (p. 3). This demands truly embracing alternative ways of being and becoming by storying our journeys with and through places (purposefully plural) "to nourish the lives of all of us who have been variously colonized" (Pierre et al., 2020, p. 407), also within our universities.

References

Ajibade, I. (2017). Can a future city enhance urban resilience and sustainability? A political ecology analysis of Eko Atlantic city, Nigeria. *International Journal of Disaster Risk Reduction, 26*, 85–92. https://doi.org/10.1016/j.ijdrr.2017.09.029

Alvarez, M. K., & Cardenas, K. (2019). Evicting slums, 'Building back better': Resiliency revanchism and disaster risk management in Manila. *International Journal of Urban and Regional Research, 43*(2), 227–249. https://doi.org/10.1111/1468-2427.12757

Augé, M. (1995). *Non-places: Introduction to an anthropology of supermodernity*. Verso.

Betteridge, B., & Webber, S. (2019). Everyday resilience, reworking, and resistance in North Jakarta's kampungs. *Environment and Planning E: Nature and Space, 2*(4), 251484861985398. https://doi.org/10.1177/2514848619853985

Brown, M., & Knopp, L. (2003). Queer cultural geographies—We're here! We're Queer! We're over there, too! In K. Anderson, M. Domosh, S. Pile, & N. Thrift (Eds.), *Handbook of cultural geography* (pp. 1–15). Sage.

Colven, E. (2017). Understanding the allure of big infrastructure: Jakarta's Great Garuda Sea Wall project. *Water Alternatives, 10*(2), 15.

Goh, K. (2019). Urban waterscapes: The hydro-politics of flooding in a sinking city. *International Journal of Urban and Regional Research, 43*(2), 250–272. https://doi.org/10.1111/1468-2427.12756

Henrique, K. P. (2023). Repositioning marginal spaces in climate adaptation: Periphery, power and possibility. In M. Armiero, E. Turhan, & S. Paolo De Rosa (Eds.), *Urban movements and climate change: Loss, damage and radical adaptation* (pp. 161–181). Amsterdam University Press. https://doi.org/10.5117/9789463726665_ch08

Henrique, K. P., & Tschakert, P. (2019). Contested grounds: Adaptation to flooding and the politics of (in)visibility in São Paulo's eastern periphery. *Geoforum, 104*, 181–192. https://doi.org/10.1016/j.geoforum.2019.04.026

Henrique, K. P., & Tschakert, P. (2022). Everyday limits to adaptation. *Oxford Open Climate Change, 2*(1), kgab013. https://doi.org/10.1093/oxfclm/kgab013

Henrique, K. P., Tschakert, P., Bourgault du Coudray, C., Horwitz, P., Krueger, K. D. C., & Wheeler, A. J. (2022). Navigating loss and value trade-offs in a changing climate. *Climate Risk Management, 35*, 100405. https://doi.org/10.1016/j.crm.2022.100405

hooks, bell. (1984). *Feminist theory from margin to center*. South End Press.

Kinkaid, E., Parikh, A., & Ranjbar, A. M. (2022). Coming of age in a straight white man's geography: Reflections on positionality and relationality as feminist anti-oppressive praxis. *Gender, Place & Culture, 29*(11), 1556–1571. https://doi.org/10.1080/0966369X.2021.2020733

Leung, M. (2025). Chapter 5: What mobility? What internationalization? Decentering for pluriversal higher education. In *Universitas* (pp. 117–136). Lexington Books. https://doi.org/10.5771/9781666952292-117

Lugones, M. (2003). *Pilgrimages/Peregrinajes: Theorizing coalition against multiple oppressions*. Rowman & Littlefield Publishers.

Manzi, M., dos S., F.,. G. C., Mourad, L. N., & de M., R.,. T. (2018). Neighbourhood planning and the right to the city: Confronting neoliberal state urban practices in Salvador, Brazil. *International Journal of Urban Sustainable Development, 10*(1), 1–15. https://doi.org/10.1080/19463138.2018.1433677

Massey, D. (2013). *Space*. John Wiley & Sons.

McConnell, F. (2017). Liminal geopolitics: The subjectivity and spatiality of diplomacy at the margins. *Transactions of the Institute of British Geographers, 42*(1), 139–152. https://doi.org/10.1111/tran.12156

Novotny, E. (2025, January 17). *Library guides: Indigenous peoples in Pennsylvania history: Towards a Penn state land acknowledgement*. Retrieved date 15 June 2025. https://guides.libraries.psu.edu/c.php?g=1145236&p=8358588.

Padawangi, R. (2014). *Humanistic planning and urban flood disaster governance in Southeast Asia: Metro Manila and Jakarta* (No. 228; Asia Research Institute Working Paper Series). Asia Research Institute.

Parikh, A. (2019a). Insider-outsider as process: Drawing as reflexive feminist methodology during fieldwork. *Cultural Geographies, 27*(3), 147447401988775. https://doi.org/10.1177/1474474019887755

Parikh, A. (2019b). Liminality of women's leisure in Mumbai, India. *Environment and Planning C: Politics and Space, 42*(5), 2399654419859358. https://doi.org/10.1177/2399654419859358

Pierre, B., Petigny, N., & Nagar, R. (2020). Embodied translations. In A. Datta, P. Hopkins, L. Johnston, E. Olson, & J. M. Silva (Eds.), *Routledge handbook of gender and feminist geographies* (1st ed., pp. 401–409). Routledge. https://doi.org/10.4324/9781315164748-41

Porto-Gonçalves, C. W. (2017). De saberes e de territórios: Diversidade e emancipação a partir da experiência latino-americana. In *Geografia e giro descolonial: Experiências, ideias e horizontes de renovação do pensamento crítico* (pp. 37–54). Letra Capital.

Pritchard, A., & Morgan, N. (2006). Hotel Babylon? Exploring hotels as liminal sites of transition and transgression. *Tourism Management, 27*(5), 762–772. https://doi.org/10.1016/j.tourman.2005.05.015

Putuma, K. (2024). *We have everything we need to start again: Empowering poems to bring you hope*. Bonnier Zaffre Ltd.

Truelove, Y. (2011). (Re-)Conceptualizing water inequality in Delhi, India through a feminist political ecology framework. *Geoforum, 42*(2), 143–152. https://doi.org/10.1016/j.geoforum.2011.01.004

8

Dwelling: Domesticity, Decay and Inhabiting Otherwise

Natalie Osborne

Abstract This chapter considers how modes of dwelling and inhabitation on stolen land are shaped by settler-colonialism, and how suburban dwelling is both troubled and constituted by non-human others and environmental change. Through a poem about a dilapidated and beloved timber house, where the white settler author dwells in collaborative and contested community with plants, critters, fungi and floodwater, this piece confronts the myths of the bordered and bounded self, and settler-colonial-patriarchal notions of mastery and control over the domestic space. Storying these intimate, everyday encounters demonstrates how self, place and Others are blurred, in a form that suggests ambivalence, uncertainty, contingency and open-endedness. Through practices of care and through lively decay, some forms of being are undone; in the undoing, other modes of living and dwelling become possible.

Keywords Dwelling · Home · Care · Decay · More-than-human

U. Kothari et al., *Stories of Place* https://doi.org/10.1007/978-981-95-4921-4_8

queenslander on Peach (Jagera and Turrbal Country)

A queenslander is a pretence of an inside

A performance of an interior

In summer, a porous membrane, exhaling

Wooden louvres painted open

Orb spiders weave flyscreens in the windows

In winter I fold her back on herself, but it's only a gesture of enclosure

She's still porous, still composting,

Regardless of season, dried leaves skitter across the floor

Under decades of decomposing frangipani leaves mycelium grows, and waits, and knits themselves

After rain, the yard fills with mushrooms and toads.

Possums don't discern between a compost bucket left ajar, and the guava tree when she is fruiting

I find bird poo and feathers in strange places, and wonder if a bust of Pallas would tempt them to stay

Plants grow in old candle jars on the windowsill, propagated from what grows below

The dishwasher drains through a hole in the floor

I'm relieved when the skitter in the corner of my eye is a skink or gecko, not anything larger or warmer

Roaches run up and down the shower drain, along the roots that weave through the pipes

They will burst one day, much like the python erupts from her skin, leaving her patterned paper for me to find in the crook of a bough

I wouldn't know she was there but for her traces

In summer storms the queenslander floats, a houseboat on an ephemeral stream

Disinterred from suburban fill, the stream leaves silt behind

on its journey to meet with *Kulpurum*

who swells and breaks banks in the eruptive joy of their reunion.

I shovel the silt off the front path into the drum that grows potatoes.

She grows, this house, in rain – doors don't open and windows don't close – and with each swell, each contraction, the glass slips a little further from the frames

Fairylights strung between the frangipani tree and the veranda mimic the warp and weft of the mycelium below

Left up too long, both house and tree have grown over the knotted strands

I tear down the vines that cling and climb up her walls

They keep coming back, first embracing, then dismantling.

---

queenslander on Peach is an open format poem storying *dwelling*, as a white settler, occupying the unceded lands of Jagera and Turrbal People in Magan-djin, so-called Brisbane, renting a house on stolen land. When I compiled this poem, I was no longer living on Peach Street. In the decade and a half I'd spent as a renter in Brisbane, this was the home I'd loved best. In 2019 I said goodbye to the queenslander on Peach Street, and embarked on a co-housing project with my partner and some friends, enabling each other to leave the rental market and practice modes of more communal living that were more consistent with our values and politics (and, frankly, more in line with our financial capacities).

But I miss her. Sometimes I think I became who I am in that house; that it is where I gained a clearer sense of who I am, and how I want to be in the world. It's where I composted some parts of myself, some old ideas, and grew new ones. It was a dwelling, a mode of dwelling, that unmade and remade me, and I remember her with gratitude and affection.

In this chapter, which I am approaching as an exegesis, I reflect first on the storying method I used to write *queenslander on Peach*. Then, I discuss two key concerns, or preoccupations, I have with/about place, that emerge in the poem. These concerns, I believe, reflect my academic work (as an urban and environmental planner and human geographer) and the grassroots work I am engaged in as a community organiser and activist on matters relating to social, spatial and environmental justice.

The first concern is an interest in unsettling my own settler modes of inhabitation, and curiosity about modes of connecting to and offering care to and with place without reinscribing settler possession (that is to say, domination) (Moreton-Robinson, 2015), and without moving to innocence (Tuck & Yang, 2012). The second is an interest in multispecies communities, particularly in how they disrupt the Western/colonial human-nature binary so frequently reproduced in Western thought and in settler-colonial land use planning. This part of my thinking is often marked with grief, anticipatory and realised, arising from the slow violence of environmental damage (Nixon, 2013; see also Dank in Radio Reversal Collective, 2025), and the "slow unravelling of intimately entangled ways of life" (van Dooren, 2014, p. 12). Illusions of (white, Western, patriarchal, capitalist) human mastery have wrought planetary damage *and* this illusion of mastery must be unlearned, dismantled, if we are to have any hope of contributing to healing the damage for which we are responsible (Singh, 2018).

Both concerns, I believe, are expressions of a desire to find and story ways to *dwell otherwise,* in, through, and with place. This desire remains incompletely understood and only met in glimpses, partially, and imperfectly, through the lens of my own compromised locatedness. Experimentation helps us cross the distance between desiring otherwise and being otherwise, so, from this compromised and contingent location, I offer some reflections on experiments. Before I explore these reflections and concerns in more detail, a few words on the method and context for writing *queenslander on Peach.*

A note of explanation: a 'queenslander' is a vernacular architecture often associated with the state of queensland (lower case used here to note that this is, of course, *not* any Queen's land!), Australia, that reflects the subtropical-tropical climate of the state (see Fig. 8.1). It is a timber frame house on stilts with a pitched tin roof, often a deep or wraparound verandah, casement windows with tin awnings, high ceilings and fretwork breezeways above internal doors (Craik, 1990; Neilsen & Pojani, 2020). These houses are designed to remain cool by allowing air circulation, so they are very open to the surrounding environment. They offer little auditory privacy from neighbours, though wooden lattice screens are sometimes used to offer some visual privacy. The freestanding, single-skin timber construction makes them amenable to DiY home renovations, so as they age, as they are lived in, inhabitants often modify them by swallowing a deck or verandah and pushing another one further out, or enclosing part or all of the undercroft for additional space (albeit compromising the cooling effect and pest and flood mitigation offered

Fig. 8.1 Street art piece of a queenslander by Phoebe Paradise, June 2025. *Source* photo by N. Osborne (2025)

by the stilted construction), with the results often ramshackle and improvised, even to the point of "sheer anarchism... where details, sizes and internal organisation run riot" (Craik, 1990, p. 197).

The term *queenslander* refers not only to this type of dwelling, but also to the dwellers. 'Queenslanders' are those who live in the state of queensland. Scholars and artists have explored the relationship between this form of dwelling and the cultures, politics, lifestyles, moods and values of those who dwell within them, and the ways these dwellings connect inhabitants to place and to post-invasion heritage (Craik, 1990; Neilsen & Pojani, 2020; see also the works of Phoebe Paradise, 2025, and the music of Rivermouth, 2021).

I did not grow up as a queenslander; I'm from further south, and spent most of my young life on the lands of the Wurundjeri Woi Wurrung and Bunurong People in Victoria, by way of Kaurna Country and Boandik Country, where I was born and where my ancestors colonised. My first home in queensland was a queenslander, and I loved it, and them, immediately. I lived in several before I moved to the queenslander on Peach Street, in a suburb in the inner south of Magan-djin/Brisbane.

I wrote this poem not as a poem at all; it was formed from fragments, lines scrawled in journals and letters written during the years I dwelled in that old queenslander on Peach Street. When the *Storying Geography Collective* decided our next project would be stories with and about place, I went back through my scraps of writing and correspondence written during my time there, to try and story that place a little. The process of going back over notes, re-arranging lines, writing some new connective tissue between ideas, images and anecdotes, gradually accreted into the somewhat clumsy open format poem that appears here. I am not an experienced or proficient writer of poetry. I enjoy reading poetry, especially with my Brisbane Free University reading group; I always feel I understand it more when it is spoken and shared in a place, in relation, and where interpretations and meanings are gathered up through conversation and shared reflection. Aside from this peer-based learning, I've little formal instruction in poetry and limited experience; when it comes to form, I have a hard time identifying what it is I like about the poems I like, and what I dislike about those I do.

All this to say, my apologies to poets and appreciators of poetry. I beg your pardon!

One of the challenges for me in practising storying with the Storying Geography Collective is overcoming my tendency to be didactic. I mostly write fairly traditional academic publications and teaching materials, where (rightfully or not!) clarity of intent is generally valued over evocation of feeling, or co-generation of meaning. I feel somewhat exposed, vulnerable, in saying much less, and being less prescriptive about my meaning (that said, this exegesis offers me a retreat to a more familiar format and offers me more explanatory space). I also do not wish to overly abstract the meanings of the feelings I was trying to express over the years I wrote the fragments that were assembled into *queenslander on Peach*—reductionism is a risk when writing an exegesis. But, as Behrendt (2016, p. 6) argues, "often there is a motivation—a politics—that accompanies the telling of any story", and this exegesis offers me a chance to reflect on that motivation and politics. It is to these reflections I now turn.

As stated above, I come to be in this place through centuries of settler-colonialism. As such, my dwelling here is fraught, marked by violence (Kotef, 2020). My dwelling here is precarious, because it is not valid. It is founded on theft, lies, malicious misconstruals, displacement, dispossession and brutality. We (settlers, white settlers) know this, viscerally (see Hage, 2002 and Moreton-Robinson, 2015, on white colonial paranoia), and so we overcompensate, telling stories of mastery and control, stories that become the mental and affective infrastructure to sustain white possession and the settler-colonial state, and to launder and justify violence against First Nations Peoples (Banivanua Mar, 2012; Behrendt, 2016; Birch, 1992; McQuire, 2024; Moreton-Robinson, 2015; Whittaker, 2018).

When I was compiling this story, I was in the process of wrapping up a collaboration called *Housing Justice in Unjust Cities* (a spin on Roy & Malson's, 2019 *Housing Justice in Unequal Cities*), where we (Anna Carlson, Jonathan Sriranganathan, Mo Chan and I) were trying to contend with the housing crisis in a way that accounted for and reflected the housing system's relationship to settler-colonialism and the material fact of Aboriginal sovereignty. As a group of left-wing community organisers, we were troubled by how the claims our movements were making for housing justice often reinscribed settler entitlements to and imaginaries of housing, and dwelling, and the possibility of a benevolent state (Carlson et al., 2024a; Porter, 2014; Porter & Kelly, 2022). In describing the many injustices and harms of the for-profit housing system, it can be too easy for non-Indigenous activists and scholars to lose sight of the relationship between the original and ongoing dispossession of First Nations Peoples, and how that model of "property-generating theft" (Nichols, 2018, p. 22) forms the basis of many resulting and ongoing injustices (Blatman & Sisson, 2024). Similarly, if we do not foreground Indigenous critiques of the settler-colonial state (see, for instance, Moreton-Robinson, 2015; Watego, 2021), we may articulate political demands that reinvest us in the maintenance of said state, rather than its undoing. As we wrote in our reflections on the *Housing Justice in Unjust Cities* project:

> By ceding the terrain of housing and spatial justice to a technocratic debate about housing policy operating on terms set by the colonial state, we overlook the possibilities offered by explicitly anti-colonial, anti-racist, abolitionist feminist theories of change that do not work to *reform* the colonial nation state, but to abolish it. The erasure of Indigenous sovereignty from these conversations narrows housing justice into a debate over who should be responsible for managing, distributing, and benefiting from stolen land, foreclosing more substantive critiques of the legitimacy of settler-colonial government itself, or the presumed 'benevolence' of the intrinsically racist colonial nation-state. (Carlson et al., 2024a, p. 19)

And yet, our group understands that solidarity between Indigenous and non-Indigenous peoples is essential for liberation (see Estes, 2019; Larsen & Johnson, 2017), and that our struggles are not the same, but our "liberation is bound up" together (Aboriginal Activists Group, 1970s, in Watson, 2004). We explored these tensions and possibilities through a series of interviews, radio shows and a public forum with a panel of Aboriginal and Torres Strait Islander speakers. This was generative, but of course yielded no easy answers, and I, we, remain uncertain about how to advance a politics of housing justice, or dwelling justice (per Porter & Kelly, 2022) that begins with and centres justice for First Nations Peoples.

This housing project was informed by an earlier community organising project around the 'right to the city', where a group of residents of Magan-djin/Brisbane organised around matters relating to public space, policing, gentrification, the housing crisis, and insurgent and participatory planning (Osborne, 2016). Some Aboriginal comrades brought to our attention the tensions and hypocrisy in our approach; while acknowledging enduring Aboriginal sovereignty over the city, we were using a Western, rights-based framework to make claims of and for it, using theories of belonging to place and spatial politics not from here, but from a colonial metropole. Why was a group of (predominately) settlers claiming yet more rights over stolen land?

I learned a great deal from these projects, not least of which was the constraints of my own habitual practices of storying and understanding place and dwelling towards more just cities. I, we, kept, often inadvertently and inconsistent with our own values, falling back into political narratives that could not accommodate the material fact of Aboriginal sovereignty. I wanted, then, to try to tell a story about place that moved towards unsettling settler modes of inhabitation and dwelling.

queenslander on Peach reflects, I believe, some of the perversities of settler modes of dwelling, like the humiliating arrogance of presuming we are masters of "our" domain, the sole beings with agency in a landscape we have only a limited and very recent understanding of. It speaks to the foolishness of modes of building cities and dwellings that ignore place, landscape, climate and variability. It reflects the ambivalence and uncertainty I felt, and continue to feel, about how to be in good relation with place. My modest hope for it is that it illustrates some of the absurdity and precarity of the myths, beliefs and stories that help sustain white possession (Moreton-Robinson, 2015), as observed from a compromised position of entanglement and complicity.

The story seeks to unsettle illusions of white-settler-colonial-heteropatriarchal notions of mastery and control over domestic space, and indeed urban space more broadly. It also reflects and subverts the (Western, colonial) myth of the bordered and bounded individual self; rather, both body and dwelling are instead permeable, ecosystemic, always vulnerable to be affected by place, by Others (Fishel, 2017; Larsen & Johnson, 2017). Precarity, uncertainty and futility mark my efforts and those who built and dwelt in the queenslander before I did to control the space. In its present state of undoing, and decay, I could not even make it an *inside*.

In the poem, we see how place re-asserts themself, resisting and refusing suburban forms. The creek that was buried when the block was first parcelled up into freehold lots (see Keenan, 2017) re-emerges, with more energy and more creative-destructive power because of attempts to hold it back. The quintessential human habitat is constantly intruded upon by plants and animals, who insist on and prefigure a place for their lives, too.

This subverts the dominant imaginary of the Australian suburban home, which is typically imagined as a site of stability and conservatism, reflecting and reproducing cisheteropatriarchy, racist hierarchies, and settler colonial and nativist belonging (Allon, 2016). The home in *queenslander on Peach* is markedly *unstable*, the structure rickety, out-of-alignment, changing with the weather. The changing climate—more frequent and more damaging flood events particularly—are accelerating this destabilisation. The dwelling is being eroded, encroached upon and into, cracks forming and widening. The structure requires constant maintenance, but we see how insufficient those efforts are against the non-human and place agencies at work. Ultimately, the dwelling is being undone, decaying, at an escalating rate. The mycelium network below is ready and waiting to redistribute what it can across the ecosystem. There's loss, but that loss is offset by the sheer vitality of this lively decay (Fishel, 2017), and it makes way for new and other forms and modes of life, of dwelling.

Decay also speaks to the temporality, and temporariness, of settler-colonialism, and its illusions of control and mastery. The agency of place (Larsen & Johnson, 2017) defies and denies the imagined permanence of the queenslander on Peach, its material form always in the process of being undone by its surroundings. As Blak activist Debbi Jones reminds us, cities remain Country: "If you listen, you can hear it through the concrete…it's right here, it's screaming at you if you can just hear it" (Jones in Brisbane Free University, 2023), and this place, this Country, asserts agency that grows cracks in the concrete. This continuity, from and to time immemorial, conflicts with settler fantasies that rely on writing ourselves into both the history and future of this place. And yet:

> Settler futures are astonishingly myopic…limited in depth, scope, and imagination. Wholesale ecological collapse, devastating wars and famines, rapid deforestation, the poisoning of vital water sources by pipelines cutting across Country, the extinction of entire species, the acidification of oceans, the aridification of once-lush plains, the leaking of oil into oceans, bushfires scorching tens of millions of hectares of bushland, the rapid heating of the entire planet, all constitute the ecological demise of a future for all. (Carlson, 2024, p. 16)

These dark white/settler imaginaries do not necessarily jolt us into care and concern; those of us raised to see (some) places and people and non-human kin as disposable may instead accelerate into nihilistic carelessness, seeking retreat or escape rather than acting on our responsibilities (Burton, 2025; Hughes, 2024). Possible antidotes for settler nihilism are, arguably, connection, solidarity and response-ability. Guided by Indigenous futurities, we may, as Carlson (2024) posits in an extraordinary collection on this topic (edited by Carlson et al., 2024b), unlearn violent settler imaginaries and learn otherwise. In struggles for and in place, through practising solidarity, we may learn how to enact care and meet our responsibilities (Larsen & Johnson, 2017), in ways that might make more just futures possible.

As mentioned earlier, we left the queenslander on Peach in order to co-house with friends. In a neighbouring suburb, together we bought an old queenslander that our housemates would reside in, now with their children, and set about building a small studio flat for my partner and I. Our love for the queenslander on Peach shaped the design for our studio. We wanted it open, permeable, full of light and air, well ventilated, breezy. We wanted it nestled in under the existing, old, mango tree, despite the issues that presented with design on an already very constrained site. We wanted to salvage materials from scrapyards, partly for environmental reasons, partly for the pleasure of learning to restore them, and partly so the battered and weathered timber that had become synonymous with home had a place in an otherwise new build.

In this new studio, *Mango Home*, possums march along the veranda and sometimes make it into the mezzanine bedroom, creeping along the headboard of the bed. A large copper skink lives under the fridge. Dried leaves still skitter across the floor. All towels must be shaken out before use because spiders love the bathroom. The window in there opens right up, making the bath almost feel like an outdoor tub, tucked under a tree fern that will, one day, be mighty. The butcher birds who visit the compost bins and the fishpond sometimes make their way inside and leave poop on the banister. The little deck built on old railway sleepers now shelters a blue tongue lizard, who scares and fascinates the dogs in equal measure. The garden is shaky, we are still repairing the damage of years of suffocation under concrete and adding more organic material back into the compacted clay, but we've added bushes and trees that are bringing more birds and insects back to the yard, and every so often it erupts in unpredictable beauty and abundance, reminding us that healing is possible, with care and patience and the unfathomable grace of the earth. And now there are children, lively and chaotic and tender and joyful, whose growth we mark against the Illawarra flame tree planted not long after the arrival of the eldest. The lowest branch is now strong enough for them to swing on, as long as they have help to reach it. I am undone over and again by all this love and liveliness.

These experiments are insufficient. I'm still occupying stolen land. It's not an innocent position; it's not a more ethical one. Comfort and security in our home life can tempt us to reinvest in the settler-colonial apparatus, which promises capital gains in exchange for compliance and orderliness. Yet, at the scale of dwelling, located in the everyday, stories and practices for dwelling otherwise may help us unsettle some ideas, help us attend to who and where we are entangled with, and teach us ways to relate to, and care for and with each other, in and with place. We practice storying and listening to stories vulnerably, in so doing "knowing that [we] are being changed and [we] are changing others" (Reardon-Smith, 2024). Perhaps, from that position of determined vulnerability, from affecting and being affected, we can unsettle and interrupt the isolation and individuality of dominant cisheteropatriarchal domesticity. We can insist on the necessity for paying more attention and caring more actively, practising response-ability with a view to unsettling dominant Western binaries of who and where matters. We can help each other better attend to our entanglements with Others as we experiment with dwelling otherwise.

In reviewing this story together, the Collective settled on the verb (though in this context, equally a noun) *Dwelling*, for how the story foregrounds modes of dwelling and inhabitation that are both troubled and constituted by non-human others, by place and by environmental change. In storying these intimate, everyday encounters, I sought to demonstrate a blurring of self, place and others, inconsistent with stories and practices of colonial mastery or dominance over place and indeed, that makes such pretensions ridiculous. Just futures worth having demand settler unlearning of these pretensions, and the destruction of the systems and practices of domination and oppression they reflect and endorse. Instead, settlers and non-Indigenous people must come into, and learn to sustain being in good relation with sovereign peoples and with Country. Storying and dwelling otherwise are small parts of this work of unlearning, dismantling and relating.

This account reflects ambivalence, uncertainty and open-endedness, which may be part of coming into good relation with place as an outsider. It is an unsettling settler story, marked with precarity and impermanence, at times an edge of unease and even danger, as the collaborative and contested community of dilapidated and poorly maintained timber and tin, plants, critters, fungi, floodwater and Country, rub up against each other to constitute a site of manifold and multiplicitous dwelling. This account is also, in part, about care, and lively decay, and how these processes and relations undo some forms of being, and that in that undoing, other modes of living and dwelling may become more possible.

References

Allon, F. (2016). At home in the suburbs: Domesticity and nation in postwar Australia. *History Australia, 11*(1), 13–36. https://doi.org/10.1080/14490854.2013.11668498

Banivanua Mar, T. (2012). Settler-colonial landscapes and narratives of possession. *Arena Journal, 37/38*, 176–198. https://arena.org.au/informit/settler-colonial-landscapes-and-narratives-of-possession/

Behrendt, L. (2016). *Finding Eliza: Power and colonial storytelling*. University of Queensland Press.

Birch, T. (1992). 'Nothing has changed': The making and unmaking of Koori Culture. *Meanjin*. https://meanjin.com.au/essays/nothing-has-changed-the-making-and-unmaking-of-koori-culture/

Blatman, N., & Sisson, A. (2024). Rethinking housing inequality and justice in a settler colonial city. In K. Jacobs, K. Flanagan, J. De Vries, & E. MacDonald (Eds.), *Research handbook on housing, the home and society* (pp. 548–566). Edward Elgar Publishing.

Brisbane Free University. (2023). Podcast 2: Underneath the car park, country, with Debbi Jones. *Forms of assembly: car park tour of inner Brisbane* [audio podcast].

Burton, A. L. (2025). Geographical imaginaries of escape: Discourses of escapism in the Tasmanian archive. *Geographical Research*. (online ahead of print). https://doi.org/10.1111/1745-5871.70008

Carlson, A., Osborne, N., Sriranganathan, J., & Chan, M. (2024a). 'The rent is too damn high' meets 'pay the rent': Practising solidarity with the dispossessed*. *International Journal of Housing Policy, 25*(3), 453–476. https://doi.org/10.1080/19491247.2024.2367837

Carlson, B. (2024). The future is Indigenous. In B. Carlson, M. Day, S. O'Sullivan, & T. Kennedy (Eds.), *The Routledge handbook of Australian Indigenous peoples and futures* (pp. 9–25). Routledge.

Carlson, B., Day, M., O'Sullivan, S., & Kennedy, T. (Eds.). (2024b). *The Routledge handbook of Australian Indigenous peoples and futures*. Routledge.

Craik, J. (1990). The cultural politics of the Queensland house. *Continuum, 3*(1), 188–213. https://doi.org/10.1080/10304319009388157

Estes, N. (2019). *Our history is the future: Standing Rock versus the Dakota Access Pipeline, and the long tradition of Indigenous resistance*. Verso Books.

Fishel, S. R. (2017). *The microbial state: Global thriving and the body politic*. University of Minnesota Press.

Hage, G. (2002). Multiculturalism and white paranoia in Australia. *Journal of Migration and Integration, 3*(3–4), 417–437. https://doi.org/10.1007/s12134-002-1023-6

Hughes, T. (2024). The political theory of techno-colonialism. *European Journal of Political Theory*. (online ahead of print). https://doi.org/10.1177/14748851241249819

Keenan, S. (2017). Smoke, curtains and mirrors: The production of race through time and title registration. *Law and Critique, 28*(1), 87–108. https://doi.org/10.1007/s10978-016-9194-z

Kotef, H. (2020). *The colonizing self: Or, home and homelessness in Israel/Palestine*. Duke University Press.

Larsen, S. C., & Johnson, J. T. (2017). *Being together in place: Indigenous coexistence in a more-than-human world*. University of Minnesota Press.

McQuire, A. (2024). *Black witness: The power of Indigenous media*. University of Queensland Press.

Moreton-Robinson, A. (2015). *The white possessive: Property, power and Indigenous sovereignty*. University of Minnesota Press.

Neilsen, V., & Pojani, D. (2020). Perspectives on built heritage preservation: A study of Queenslander homeowners in Brisbane, Australia. *Journal of Housing and the Built Environment, 35*, 1055–1077. https://doi.org/10.1007/s10901-020-09767-z

Nichols, R. (2018). Theft is Property! The recursive logic of dispossession. *Political Theory, 46*(1), 3–28. https://doi.org/10.1177/0090591717701709

Nixon, R. (2013). *Slow violence and the environmentalism of the poor*. Harvard University Press.

Osborne, N. (2016). *The right to the city: In pursuit of a more just, sustainable and democratic Brisbane*. Policy Innovation Hub. https://medium.com/the-machinery-of-government/the-right-to-the-city-a4ffb9d4dc76

Phoebe Paradise. (2025). *Phoebe Paradise.* Retrieved date March 2025. https://www.phoebeparadise.com/

Porter, L. (2014). Possessory politics and the conceit of procedure: Exposing the cost of rights under conditions of dispossession. *Planning Theory, 13*(4), 387–406. https://doi.org/10.1177/1473095214524569

Porter, L., & Kelly, D. (2022). Dwelling justice: Locating settler relations in research and activism on stolen land. *International Journal of Housing Policy, 23*(4), 817–835. https://doi.org/10.1080/19491247.2022.2132461

Radio Reversal Collective (2025). *Caring for country against the colonial polycrisis with Rikki Dank [audio podcast]*. Retrieved date March 2025. https://soundcloud.com/radioreversalcollective/episode-15-caring-for-country-against-the-colonial-polycrisis-with-rikki-dank?utm_source=clipboard&utm_medium=text&utm_campaign=social_sharing

Reardon-Smith, H. (2024). Musickin as in Free Palestine: Vulnerable listening in violent times. *Disclaimer*. https://disclaimer.org.au/contents/vulnerable-listening

Rivermouth (2021). *Subtropical metropolis? [album]*. Retrieved date March 2025. https://rivermouthband.bandcamp.com/album/subtropical-metropolis

Roy, A., & Malson, H. (Eds.). (2019). *Housing justice in unequal cities*. University of California. https://escholarship.org/uc/item/4kq1j0df

Singh, J. (2018). *Unthinking mastery: Dehumanism and decolonial entanglements*. Duke University Press.

Tuck, E., & Yang, K. W. (2012). Decolonization is not a metaphor. *Decolonization: Indigeneity. Education & Society, 1*(1), 1–40. https://clas.osu.edu/sites/clas.osu.edu/files/Tuck%20and%20Yang%202012%20Decolonization%20is%20not%20a%20metaphor.pdf

van Dooren, T. (2014). *Flightways: Life and loss at the edge of extinction*. Columbia University Press.

Watego, C. (2021). *Another day in the colony*. University of Queensland Press.

Watson, L. (2004). *"Recognition of Indigenous Terms of Reference," Keynote Address at "A Contribution to Change: Cooperation Out of Conflict Conference: Celebrating Difference, Embracing Equality," Hobart*, September 21–24, 2004. Retrieved date March 2025. https://uniting.church/lilla-watson-let-us-work-together/

Whittaker, A. (2018). *Blakwork*. Magabala Books.

9

Enchanting: Homage to My Mum and Sandhills Community Garden

Phoebe Everingham

Abstract People story places and places story people. Places are more than what can be expressed through words; they contain ambience, and multilayered, multitemporal traces that affect and are affecting. This chapter explores the site of Sandhills Community garden, a place that sits on the unceded lands of Awabakal Country in so-called Australia. Layers of colonial violence are entangled within everyday affective bonds of joy and care that arise from more-than-human moments of relational liveliness. Here, I story the ambivalence of the extraordinary within such familiar and ordinary places. Drawing on geographies of enchantment I consider the role of emotions such as wonder, awe, grief, nostalgia and melancholia to consider the worldmaking potentials of stories. While Sandhills Community garden highlights the contested nature of place, I explore the possibilities for how everyday extraordinary affects may forge new emotional bonds between people and place.

Keywords Enchantment · Hope · Affect · Community garden

U. Kothari et al., *Stories of Place*, https://doi.org/10.1007/978-981-95-4921-4_9

Fig. 9.1 The first stages of building the garden, preparing the garden beds. *Source* Christine Everingham's personal photo

Fig. 9.2 A hand-painted sign that Jen made with the children. *Source* Christine Everingham's personal photo

A shift worker sits at the picnic table in the early morning, glad to have found a peaceful place to wind down and read the paper before going home. An office worker spends her lunchtime in the garden to 'rejuvenate' before going back to work. A case worker brings her client from an outer suburb to visit—this is an exciting 'outing' across town and the garden brings them both peace. A busload of senior citizens pull up—they wander through, full of nostalgia, reminiscing about their gardens they used to tend, that they no longer have, since moving into the retirement home. An older lady with a motorised scooter arrives in the garden, her little dog sitting on her lap. She places the dog carefully on the ground and it runs off for a play and a sniff—she needs a peaceful safe place to go where her dog can run around while she watches on fondly. She also hopes there is someone in the garden she can chat to about her family problems. Others come and go, some for the first time wanting to know what grows there, what is in bloom, what is edible, who is responsible for the garden and how did it get here?

Twenty or more years ago after the death of my father my mother left our suburban family home with its beautifully tended and layered garden, frog pond and vegetable patch. She moved into an inner city terrace that gave her access to a unique beach and an inner city lifestyle; however, it had no space for a garden. A long-time community activist and organiser, she got together with a few residents to plan a community garden. An application for the use of public land was put to council and while they were waiting, over 6 months, for a yes or a no, and worried the group would lose momentum, Mum suggested they begin to clear the embankment in preparation for sowing new life. Eventually, the go ahead was given by council and a small grant of a few thousand dollars enabled them to buy some materials to begin work. Located on the site of an old power station that had made the topsoil toxic, meant that raised garden beds were the first things needed. Working bees ensued for the first 12 years once a month, and various workshops were run by permaculture enthusiasts. In the first few months people would bring beautiful materials such as sandstone, old railway sleepers and logs, to terrace the embankment—providing places to plant fruit trees for shade.

The garden has also been a place for my mum to sit with her grief, after losing her soulmate, and a place that brings joy through care, creativity and connection. She is not the only one to find solace in her grief here. A woman tells us she buried her cat under the tree—just over there. A mother who lost her son rejoices in the smell of lavender, which reminds her of him. The garden also has a life of its own—the comings and goings of people who pass through when my mother is not there planting or watering. Some people from the neighbourhood come down and plant seedlings and new life springs up in unexpected places. A fairy garden starts to take shape—a local woman who loves to bring joy to the children who visit, painted wood stakes, and carefully placed fairy statues and baubles around the garden for children to find. She gathered local children for sign painting days that contain sayings such as 'all gardens grow first in the heart' and the children place these signs in secret nooks feeling special that their sign now belongs in the garden. They too now have a sense of belonging to this place. Local gardeners claim a section to work on—they come with a vision of planting natives, to attract certain insects and birds. Others want to experiment with avocado seeds—see if they grow into trees. Just the other day we noticed some bee houses that the local pre-schoolers had placed around the garden—hoping to revitalise the dwindling bee population. Secret paths begin to take shape—a way to keep the embankment stable. The paths are demarcated using remnants of materials that Mum finds about the place to hold up the plants that line the paths—logs, branches, even an old gate frame! The paths offer sites of wonder and enchantment for the kids who like to explore while their parents sit in the sun and relax.

When my mother was first searching for a place for a garden, she walked through this part of the park. At the time it was an unused little corner of a bigger public park, its industrial history apparent in the gravel and landfill. It was sheltered by an embankment mostly overgrown with bitou bush—an introduced weed from South Africa, demonstrating the latest layer of colonial territorialisation. Rubbish was strewn in various corners and rats had made their home there. Despite this 'ugliness', Mum saw its potential and was affected by an overwhelming sense of care and custodianship. She found the space quiet and serene—it was north-facing which meant it captured just the right amount of sun. She was affected by the ambience it offered. As the garden began to take shape, we started to learn more about this place, rolling Sandhills that would have provided a beautiful campsite for Awabakal people who spent tens of thousands of years living on these lands. Not far from the banks of the river and the beach coastline was a freshwater spring.

Pre-colonisation, Newcastle was known as Moolimbimba—a place of the sea fern which grew in abundance in the river estuary. Now the river is a harbour, and there is a kilometre or more of reclaimed land where the foreshore park stands. There is no more sea fern, nor abundance of shellfish and life that once fed and nourished the people who cared for this place. People who were turfed off their lands once it was discovered Moolimbimba contained valuable coal, an important resource to keep the colony expanding, setting the stage for Newcastle to be the world's largest export site for coal, furthering the postcolonial processes of extraction, expansion and environmental injustice. The effects of the emission from the burning of coal extracted from this place are causing tides to rise, threatening the sovereignty of our Pacific island neighbours to sustain their livelihoods and culture. Fossil fuels are causing death and extinction of a myriad of non-human beings—what is to become of our future generations?

Settlement on this place happened very quickly—the river frontage was granted to wealthy landholders from Sydney. Awabakal people were locked out of access to the river, and to the Sandhills—which were paved over by a convict lumber yard—a place of hard work, discipline and torture of the convicts brought up from Sydney to play their part in further colonising the lands. The lumber yard is now considered one of Newcastle's most haunted places—a place where some claim they have felt the pinch or flick on their back as they pass through the yard in the night. This is the ghost of Robert Young, whose job it was between 1814 and 1818 to inflict more than 13,000 strokes with his cat-o-nine tails on 275 disobedient convicts and the sex workers the colony depended on.

The place where Sandhills Community garden now sits is one of dispossession and brutality. The layers of violence inflicted not only dispossessed traditional custodians of their land, but also perpetuated colonial expansion through the exploitation of convicts and settler women. However, the once peaceful place nestled on the banks of the estuary has been regenerated through new entanglements and connections at the site of Sandhills Community garden. The future ancestors of settlers, convicts, Awabakal and other nations can find peace and solace from grief in all its forms, within this place. New migrants also find ways to connect with the garden and cross-cultural exchanges occur through sharing knowledge about plants, trees and herbs. An Indian family tells us about how to use the curry leaves. A Filipino neighbour plants banana trees—and then shows us how to use the flower in a curry. We never knew this part of the plant could be eaten. Another migrant family comes once a year to cut off parts of the sugar cane—everyone gets involved, grandparents to children—and off they go to the nearby picnic table.

Sandhills Community garden highlights the contested nature of place, but also enables connections and reflects a desire to create and maintain a place of enchantment, joy and care. It contains layers of grief from the violent processes of colonisation. Yet it also disrupts this legacy and the empty individualism that such colonial capitalist processes have perpetuated through connecting people in and with place in new ways. Through these new affective and emotional bonds generations connect across cultures and within spiritual realms to forge post-colonial and post-capitalist repair and regeneration, towards futures otherwise.

In times of increasing ecological crises and social inequality it is easy to feel disenchanted, overwhelmed by feelings of sadness, anxiety, fear and mourning. Climate crises are no longer an abstract concept, but local, personal and deeply unsettling. Ecological crises have multisensory bodily impacts; the sun no longer feels sultry, the sound of the rain on the tin roof no longer feels cosy, but brings fear of bushfires or flooding. In the 250 years since Europeans colonised Australia, native birdlife has disappeared across the continent. For the regent honeyeaters, a population crash to fewer than 300 has caused the species' song culture to break down. Fragmentation of bird populations results in young birds learning songs of different more dominant species instead of their own. Their soft warbling songs are changing, impacting their ability to mate and bring new life into the world (Crates et al., 2021). These are just some examples of disenchantment in the modern world. Loss, despair and feelings of hopelessness are the affective underpinnings of "the anxiety of the Anthropo-Capitalocene" (Margulies, 2022, p. 257).

Disenchantment is linked to feelings of melancholia and a nostalgia for a time when there was a sense of wholeness, harmony with nature, connection to community and the simplicity of living with rhythms of time that aligned with sunrises and sunsets. Max Weber attributed the notion of disenchantment to the processes that threaten our attachments to these natural rhythms and flows, where rationality and clock time render the world an existentially dreary place. While the disenchantment of the modern world has led to some decrying 'the loss of a golden age', those who believe in progress models of modernity have celebrated it as "the end of a dark age and the dawn of a world of human freedom and rational agency" (Bennett, 1997, p. 1). Regardless of whether modernity leads one to nostalgic longings or feelings of transcendent liberations, the term enchantment itself underpins something that used to exist in a distant past, even if in the realms of magic and wonder. In defining the term, Jane Bennett (1997, p. 1) says:

> Just what is this enchantment from which we are purported to be dissed? 'Enchant' is linked to the verb to sing: chanter. To en-chant: to surround with song or incantation; hence, to cast a spell, to bewitch with sounds, to make fall under the sway of magical powers. A dis-enchanted world, then, would be inert and songless, unable to induce a swoon or to conjure the strange and wonderful from out of the banal and ordinary.

Bennett (cited in Krøijer & Rubow, 2022, p. 377) goes on to challenge the premise of the modern world as a 'disenchanted' one. Instead, she (Bennett cited in Krøijer & Rubow, 2022, p. 377) argues for a broadening of the scope of enchantment, which endures in the form of "fullness, plenitude, and liveliness associated with wonder and 'energetic love of the world' while keeping in mind the uncanny and disruptive qualities involved in any intense engagement with the world". To remain enchanted by and in the world requires an attentiveness to everyday moments of flourishing and liveliness, of beauty and wonder. In Bennett's framing of the term, to be enchanted is not to be falling to your knees awestruck, instead it exists within the everyday moments of affective encounters and "active engagement with objects of sensuous experience; it is a state of interactive fascination…" (Krøijer & Rubow, 2022, p. 377).

The loss of so many species in this sixth mass extinction is the canary in the coal mine of total ecosystem collapse. Amidst these feelings of overwhelm, how do we repair this damage and prefigure more caring and enchanted worlds? To care is to understand we are in a very concerning situation and to be alert to the immediacy of disruption, disjuncture, imbalance or pain (Gellek, 2018). What is the relationship between care and enchantment? Care is embedded within mundane moments of maintenance and repair that sustain everyday life, comprising everything we do to live in the world as well as possible (Puig de la Bellacasa, 2017). Krøijer and Rubow (2022) point out that a disposition towards 'caring for' rather than 'caring about' is important to understand enchantment as an ethics of care because it "nuances the spectrum between detached attention, acts of being attentive to others, and caring for others" (p. 379). To be enchanted is to be affected and attentive to mutual modes of care and 'never a human-only matter'. To link enchantment to care involves 'a choreography of multiple, overlapping practices', an openness to new and other ways of being 'we' (Krøijer & Rubow, 2022, p. 379). In this way, one can be attuned to the enchantments of Sandhills Community garden through being attentive to the more-than-human co-becoming and possibilities contained in moments of beauty and connection. Baker (2022, pp. 397–398) states:

> We all need enchantment in this time of global existential crisis and a 'counterstory' [that] seeks to induce an experience of the contemporary world—a world of inequity, racism, pollution, poverty, violence of all kinds—as also enchanted—not a tale of re-enchantment but one that calls attention to magical sites already here.

Pyyry and Aiava (2020) see enchantment as an affective event, not only of pleasant experience but also of negative registers. Affect circulates through despair, loss and suffering, nostalgia, melancholia, wonder, awe and life-affirming moments, unfolding the political and ethical potentials of enchantment, a reordering of the senses to consider how things could be otherwise. Despite the fear and overwhelm, I'm still attuned to the magical sites amongst the remnants of the ruins—the family of five tawny frog mouths that live in the tree between Sandhills Community garden and my house—we see them from my son's bedroom window. We hear them hoot when we take the wheelie garbage bin to the corner, containing the materiality of a modern-day life of overconsumption destined to landfill. Sandhills Community garden provides a space of refuge amongst the chaos, fragmentation and isolated individualism of modern life.

When thinking about how to begin this chapter and my own enchantment with Sandhills Community garden, I wandered through one morning to give me some inspiration. It's a place I visit regularly as it is right next door to my house. In fact, just today my mother has been to Bunnings hardware store and left some sawdust and manure in the front section of my gate—ready for her late afternoon gardening stint, when she will pop over to say hello in her gardening clothes and hat, before she carries her goods to begin the next pottering session. Sometimes when I drive my son to school, we laugh at the lingering smell of Grandma's manure in the car we share—we know that she has a plan in place to create a new garden bed, or mix the manure in with the soil to spread to the places she knows seedlings need extra support to thrive. She knows exactly where it is needed because of her intimate relationship with all the secret nooks and crannies, where plants sometimes emerge of their own accord—giving rise to a gasp of surprise at unexpected emergences—a pumpkin! A zucchini! A new pocket of rocket. When I stepped into the garden the other morning, I was surprised to see a smattering of new colour. The cosmos flowers had self-seeded and every year appearing out of nowhere—seeming to welcome visitors to the garden through their vibrant colours of purple and blue.

This may be what Bennett (1997, p. 7) refers to as 'chance beauty', the kind of beauty that arises through "the pleasing copresence of sensuous complexity and arranged form". Jen, the woman who paints signs with the children and carefully places objects around the garden to bring colour and novelty, had 'arranged' an old bicycle in a garden bed, with a small bunch of cosmos planted in the basket of the bike as decoration. Each year they have multiplied, and the seeds have flown around and planted themselves in places where they seem to know would be a good place to spring up and welcome people into the garden. For Bennett (1997), kaleidoscopes enchant because they mix artful design and the random juxtaposition of shapes and colours. Likewise, the cosmos flowers contain an element of unpredictability that resists a fully rational accounting for, surprising and confounding the human agency that began with a purposeful containment of cosmos in the bicycle basket.

In creating this fairy element to the garden, enchanting children and adults alike, Jen creates magic. For Bennett (1997, p. 8), magic "enables a surprising reconfiguration of the newly released bits. Those bits may recombine with each other or with the disaggregated fragments of other (former) wholes". In other words, there is a hybrid notion to magic, an experimental mixing of elements... products not amenable to established categorical distinction (Bennett, 1997, p. 8). Magic is present in the spatial higgledy piggledy layout of the garden, a setting of relational hybridity of organisation and human agency, and the uncontainability of plants, animals and more-than-human encounters.

Why does the garden that sits in the corner of a much larger park attract so many people who picnic, fossick and explore—in a way that other sections of the park do not? The bordered fixture of a council parkland seems uninviting, rigid, soulless and non-magical. Perhaps it is because these borders represent the iron cage of modernity—a feeling of being trapped, contained and stifled within rigid structures, rules and regulations, a space that does not allow for spontaneity and wonder. In contrast, Sandhills garden invites serendipitous encounters in a way the other sections of the park do not. The city is not merely a contained space where encounters occur; instead, it is made from the relationalities of multiple encounters that make spaces non-fixed, but always under construction (Wilson, 2017). For Wilson (2017, p. 453) it is these chance encounters that "give urban life its 'distinctive character', liveliness and risk, making the urban a site of 'permanent disequilibrium' where 'normalities and constraints' are continuously dissolved". As Steve Pile (2021, p. 244) writes: "The affective and emotional intensities of human and nonhuman relationships must be seen as fundamental to how and why physical structures take the form they do". Urban design choices influence emotional and affective moods, where experiences of place are influenced by bodily experiences. In this way Sandhills

Community garden provides an enchanting feeling of affective wonder and belonging in a way that the bigger foreshore park does not.

What does this 'small story' about a seemingly insignificant community garden in an urban setting tell us more broadly about our relationship to the natural world, about the ambivalences of what it might mean to 're-enchant', or to be attuned to already existing everyday enchantments? To feel affected by enchantment does not mean a negation of the layers of despair, nor the violent palimpsest of coloniality that underpins modern-day crises. Situated on the settler colonial lands of Moolimbimba, in Australia, Sandhills Community garden also contains unsettling moods and contradictory affects produced by colonialism and ecological change. While Sandhills facilitates encounters of care and connection to bigger environmental and social ecologies it is also a site of disrepair—where rolling sand dunes no longer reach to the riverbed, where the seafern no longer grows. Instead, we hear the sound of a foghorn, as the latest coal carrier rolls into the harbour, while Fort Scratchly looms on the headland, a fort built to ward off the Japanese military in World War II. From Sandhills Community garden we can hear the daily shooting of the 1 pm cannon, reminding us of the ever-present military governance that underpins these settler colonial lands.

Everyday enchantments are life-affirming; they do not negate the despair. For Pyyry and Aiava (2020), enchantment need not necessarily feel joyful or energetic; it provokes both wonder and unease. Enchantment is an encounter with playfulness, opening up sensory receptivity towards engendering care for the world. In this way, the affective moods of enchantment are not separate from ethical and political concerns. As Sebastian Ureta (2016) writes, affective entanglements are always unpredictable and chaotic, and care is never free from unsettling manifestations. Margulies (2022) poses the question: "[w]hat structures this distance between handwringing and the active work necessary for fostering life"? Care is a transformative process in which "carers become affected by the entities we care for and as such "become matters of care to us" (Di Giminiani, 2022, p. 420).

Sandhills Community garden provides layers of healing, repair and regeneration. The garden gives spaciousness to grief, my mum's loss of her soulmate (my father), the dead cat buried under the tree, ecological grief for a changing world and grief of and for Awabakal peoples whose land was stolen. Land theft has robbed subsequent settlers of an opportunity to learn how Awabakal people's care and custodianship of lands would in turn care for us. Contained in this grief is the nostalgic longings for the Sandhills that would have rolled out to the river. There is an ambience of a haunted past and yet while humans are central to the story of how the garden came to be, the more-than-human entanglements hold these complexities through anarchic agency and creativity. Within the scent of lavender, grief, joy, beauty and wonder co-exist—a symbol of life that also acknowledges death. Sandhills Community garden also tells a story of what it means to belong on settler colonial lands. It's a story that subverts notions of containment and private property because it's a space that invites openness towards spontaneous encounters of care and connection. It's an experimental site reconfiguring dominant settler ways of relating to land. In foregrounding care and connection in the garden I chose not to tell the stories of stolen plants, the argument I had with the woman I caught cutting down sunflowers to take back to her house, and the graffiti on the handmade wooden bench. The garden is a site where these contestations over space, stolen lands, sharing and caring all take place simultaneously, where these relationalities cannot be quantified into inputs and outputs, or confined to the iron cage of modernity. It is precisely because of the uncontainability of these layers of ambivalence that makes Sandhills community garden enchanting. Enchantment is found in the small everyday moments of hope and of affirmative visions for the future. Enchantment embraces the potential to become otherwise, towards shared ontological commitments and collective ways of being in the world (Wilson, 2017). Indeed, such small stories matter.

References

Baker, J. M. (2022). Boreal plants that enchant: A lively ethnography of Sakâwiyiniwak (Northern Bush Cree) multispecies kinship obligations. *Environmental Humanities, 14*(2), 385–400.

Bennett, J. (1997). The enchanted world of modernity: Paracelsus, Kant, and Deleuze. *Journal for Cultural Research, 1*(1), 1–23.

Crates, R., Stojanovic, D, Langmore, N, & Heinsohn, R. (2021). Only the lonely: An endangered bird is forgetting its song as the species dies out. *The Conversation*. https://theconversation.com/only-the-lonely-an-endangered-bird-is-forgetting-its-song-as-the-species-dies-out-156950

Di Giminiani, P. (2022). The limits of care: Vitality, enchantment, and emergent environmental ethics among the Mapuche people. *Environmental Humanities, 14*(2), 419–437.

Gellek, T. (2018). *How to create a more caring world*. Retrieved December 19, 2025, from https://www.dailygood.org/story/2131/how-to-create-a-more-caring-world-tsering-gellek/.

Krøijer, S., & Rubow, C. (2022). Introduction: Enchanted ecologies and ethics of care. *Environmental Humanities, 14*(2), 375–384.

Margulies, J. (2022). A political ecology of desire. *Environmental Humanities, 14*(2), 241–264.

Pile, S. (2021). *Bodies, affects, politics: The clash of bodily regimes*. John Wiley & Sons.

Puig de la Bellacasa, M. (2017). *Matters of care: Speculative ethics in more than human worlds*. Minnesota University Press.

Pyyry, N., & Aiava, R. (2020). Enchantment as fundamental encounter: Wonder and the radical reordering of subject/world. *Cultural Geographies, 27*(4), 581–595.

Ureta, S. (2016). Caring for waste: Handling tailings in a Chilean copper mine. *Environment and Planning A, 48*(8), 1532–1548.

Wilson, J. (2017). *Neoliberalism*. Routledge.

10

Inviting: An Epilogue

Storying Geography Collective

Abstract Briefly summarising some of the practices supporting the storytelling contained in this volume, this epilogue invites readers to reflect on what it means to story well and to story what matters for and with place. Our own work has been guided by imperatives to story differently, together, carefully, responsibly and relationally. We offer these reflections to support the storying practices of others.

Keywords care · relationality · politics · storying

We end this book, not with a conclusion, but with an opening, an invitation. Through writing, sharing, reading and thinking with each other's stories we have a strong sense that our stories continue—continue to evoke affect, generate new ideas, emotions, actions and possibilities. And so, with this book, we wish to gesture towards the possibilities of stories and invite others to engage with stories of place.

To story differently. As captured in this book, there are many ways to engage with the stories that we write, tell and encounter. Our storying practice reflects this diversity, as each chapter holds and weaves stories differently with other kinds of texts. This book is neither an edited

U. Kothari et al., *Stories of Place*, https://doi.org/10.1007/978-981-95-4921-4_10

collection nor a monograph but brings together individual writing and reflection, with collective sharing, listening, reviewing and responding.

To story together. In our collective we have approached storying as a practice carried out together, in relation to others, with place and with the non-human. Working together in a supportive way gave each of us a sense of safety, a space to be vulnerable, to know our stories are worth sharing and to be confident in experimenting with the practice of storytelling.

To story carefully. We have tried to approach the practice of creating, writing and recounting stories in ethical ways and with care about places and relations. We acknowledge the troubled history and practice of storytelling and the ways it is purposefully used to obfuscate, oppress and silence the histories, knowledges and perspectives of those from the margins or whose attachments to place disrupt dominant interests. We recognise that stories and storytelling are inherently political, yet, when engaged with in critically reflexive and ethical ways stories have the potential to be emancipatory and transformative, contributing in consequential ways to enacting other worlds.

To story responsibly. Feminist and anti-colonial critiques of Western knowledge production demonstrate ways of practising knowledge (co-)production that dissolve established binaries between body/mind and reason/emotion. This is especially so when adopting a relational, ethical and collective practice; one that has led us to the joy and generative possibilities of stories. In this time of polycrisis, we need particular kinds of stories—stories that are nurturing, that evoke feelings that contribute to change, to connect us to each other, to place and the living world. Stories are shared often in relation to others and express people's responsibilities to others—to the human, to the more-than-human, to place, to memories. It is important to honour such connections and relations, listen to others' stories, including more-than-human stories, acknowledging the agency of the listener. Yet, we also respect silences and refusals, not all stories are for sharing.

To story relationally. We have practised relational storytelling in ways that celebrate difference. We have not shied away from storying about the everyday, seemingly mundane places that give shape and meaning to our lives—houses, gardens, potholes, cemeteries, picnic spots, buildings, offices, bursaries. All are extraordinary because of the relations we have with those places. Storying is thus a way to honour the significance of something that might be overlooked or forgotten.

To story well. The stories we have shared speak to places as agential, embodied, relational, emergent and co-produced, recognising that storytelling is important to identity, community, connection, politics and memory. In *Storying Place*, we have tried to challenge hegemonic narratives of neoliberalism, unfettered economic growth, whiteness, patriarchal and racialised capitalism, colonialism and environmental destruction. Part of taking storying seriously is to look at how a story could be considered ethical and meaningful; thoughtfully and engagingly told yet capable of upholding an author's intention. We are concerned with the circulation and amplification of particular stories, and the extent to which stories contest, challenge, transform, or reinforce and reinscribe systems of social-spatial domination.

To story what matters. We respond to the power of stories particularly at this time when many struggle to make sense of their lives, in the face of ongoing and intensifying ecological harm, inequality, violence, oppression and displacement. It is important to consider then, which places are storied, how and when they are storied, and by whom, and to illuminate the power of stories to make and unmake worlds. We recognise that stories can be ambiguous, liminal, mobile, imagined and bordered. This requires being attentive to the kinds of stories we tell and imbibe, which ones come to dominate, and which ones are shaping our imaginations, aspirations and actions, enabling or disabling possibilities.

We invite others to tell stories as if place matters.

Index

U. Kothari et al., *Stories of Place*, https://doi.org/10.1007/978-981-95-4921-4

FSC
www.fsc.org
MIX
Papier | Fördert gute Waldnutzung
FSC® C083411

Kolibri 360 GmbH
Ferdinand-Jühlke-Straße 7
99095 Erfurt, Deutschland
produktsicherheit@kolibri360.de